GEOGRAPHICAL READINGS

Glaciers and Glacial Erosion

The Geographical Readings series

Published

Rivers and River Terraces G. H. DURY

Introduction to Coastline Development J. A. STEERS

Applied Coastal Geomorphology J. A. STEERS

World Vegetation Types S. R. EYRE

Developing the Underdeveloped Countries ALAN B. MOUNTJOY

Glaciers and Glacial Erosion CLIFFORD EMBLETON

Titles include in preparation

Climatic Geomorphology E. DERBYSHIRE

Biogeography R. P. MOSS

Transport and Development B. S. HOYLE

English Rural Communities D. R. MILLS

Glaciers and Glacial Erosion

EDITED BY

CLIFFORD EMBLETON

MACMILLAN

First published 1972 by
THE MACMILLAN PRESS LTD
London and Basingstoke
Associated companies in New York Toronto Dublin Melbourne Johannesburg and Madras

SBN 333 12655 6 (hard cover)
333 12656 4 (paper cover)

Printed in Great Britain at
THE PITMAN PRESS
Bath

Contents

Acknowledgements

'Glacial Erosion in France, Switzerland and Norway', by William Morris Davis, from *Proceedings of the Boston Society of Natural History*, XXIX (1900) 273–321, by permission of the Museum of Science, Boston, Massachusetts

'The Profile of Maturity in Alpine Glacial Erosion', by Willard D. Johnson, from *Journal of Geology*, XII (1904) 569–78, by permission of the University of Chicago Press

'Crescentic Gouges on Glaciated Surfaces', by G. K. Gilbert, from *Bulletin of the Geological Society of America*, XVII (1906) 303–13, by permission of the Geological Society of America

'Geologic History of the Yosemite Valley', by François E. Matthes, from *Professional Paper 160 of the U.S. Geological Survey* (1930) 54–103, by permission of the U.S. Geological Survey

'Observations on the Drainage and Rates of Denudation in the Hoffellsjökull District', by Sigurdur Thorarinsson, from *Geografiska Annaler*, XXI (1939) 189–215, by permission of the Swedish Geographical Society

'Radiating Valleys in Glaciated Lands', by David L. Linton, reprinted from *Tijdschrift van het Koninklijke Nederlandsche Aardrijkskundig Genootschap*, LXXIV (1957) 297–312, by permission of Geografisch Tijdschrift

'The Forms of Glacial Erosion', by David L. Linton, reprinted from the *Transactions of the Institute of British Geographers*, XXXIII (1963) 1–27, by permission of the Institute of British Geographers

'Glacial Erosion in the Finger Lakes Region, New York State', by K. M. Clayton, from *Zeitschrift für Geomorphologie*, IX (1965) 50–62, by permission of Professor Clayton and Gebrüder Borntraeger Verlagsbuchhandlung

'Measurements of Side-slip at Austerdalsbreen, 1959', by J. W. Glen and W. V. Lewis, reprinted from the *Journal of Glaciology*, III 30 (1961) 1109–22, by permission of the Glaciological Society

'The Flow Characteristics of a Cirque Glacier and their Effect on Cirque Formation', by J. G. McCall, from W. V. Lewis (ed.), *Investigations on Norwegian Cirque Glaciers*, Royal Geographical Society Research Series, IV (1960) 39–62, by permission of the Royal Geographical Society

'Direct Observation of the Mechanism of Glacier Sliding over Bedrock', by Barclay Kamb and E. LaChapelle, reprinted from the *Journal of Glaciology*, V 38 (1964) 159–72, by permission of the Glaciological Society

'The Theory of Glacier Sliding', by J. Weertman, reprinted from the *Journal of Glaciology*, V 29 (1964) 287–303, by permission of the Glaciological Society

Introduction

THE movement of glaciers is a phenomenon that has been known for more than four centuries, and the fact that ice in the world once covered areas vastly greater than it does at present has been widely accepted for over 100 years. Gradually, it has come to be recognised that moving ice is one of the major sculptors of the earth's surface; indeed, evidence is accumulating to suggest that, quantitatively, it may be the most potent agent of terrestrial erosion. The papers and extracts in this anthology have been selected to span the period beginning with some of the earliest pioneer investigations into glacier movement and glacier erosion processes, and leading up to the research of the last two decades in which great advances in our knowledge of those processes have been achieved.

The balance of papers is roughly equal between old and new. It is important that students of any science are as aware of the older 'founding' contributions to the literature as of the modern developments and that they should be familiar with the 'historical weapons and arguments'. 'Every generation enjoys the use of the vast hoard bequeathed to it by antiquity, and transmits that hoard, augmented by fresh acquisitions, to the future ages. In these pursuits, therefore, the first speculators lie under great disadvantages, and when they fail, are entitled to praise' (Macaulay, quoted by J. K. Charlesworth, 1957). The choice of papers from the vast body of literature on glacial geomorphology is not an easy one. I have attempted to include papers of widely differing approaches, including examples of the deductive approach on the one hand and the quantitative approach on the other. Some of the papers are now difficult to obtain or are to be found in periodicals not well known to most students. The choice of 'modern' papers was perhaps the most difficult; whereas one can survey the older literature from the standpoint of present knowledge and discern what have turned out to be the most significant contributions, it is not easy to predict with any certainty which of the papers published over the last decade will be judged significant in another fifty years. The papers that I have selected from those published in the

1960s are not necessarily the most 'up to date'; many views expressed in them would doubtless be modified if the authors were writing those papers today, and, indeed, more recent papers by the same authors on related topics have appeared. But each of the ones selected in this anthology stands as a milestone marking progress in its particular field, and they are all, without question, 'required reading' for students of glacial geomorphology. It is most important to note that, owing to limitations of space, some papers have had to be drastically shortened and all photographs have been omitted. No indication is made of the places where sentences, paragraphs or whole sections have been omitted. This book should, therefore, not be used as a source of reference to original material, or for purposes of quoting an author's original text. However, most of the changes have consisted solely of the omission of complete paragraphs or sections, and the fewest possible changes have been made to the phraseology, so that the flavour of the original style has been maintained.

As the Glacial Theory had its beginnings in the Alps, it is appropriate to begin with some extracts from that fascinating pioneer work by James D. Forbes, *Travels Through the Alps of Savoy*, published in 1843. Forbes's measurements of the rate of motion of the Mer de Glace represented one of the first comprehensive studies of glacier motion, and a model for future workers. Two years previously, L. Agassiz had shown that the centre of the Unteraar glacier flowed faster than its sides. Forbes in 1842 carried out a much more complete analysis of glacier surface motion. The results that Agassiz and Forbes obtained with the primitive instruments at their disposal are remarkable, and their deductions as to the nature of glacier motion were far ahead of their time. Forbes distinguished correctly between internal deformation of the ice and basal sliding; he concluded, also correctly in the case of temperate glaciers, that movement varied with the state of the weather and between summer and winter; and his deduction that movement diminished steadily from the ice surface to the glacier bed was shown to be substantially correct by borehole experiments carried out a century later by M. Perutz and subsequent workers. Forbes also suggested, with great perspicacity, that whereas some glaciers might slide over their beds, others in colder conditions might not if they were frozen to bedrock. The extract from his book reproduced here describes his measurements on the Mer de Glace, and his deductions on the motion of glaciers. Also included is a short account from another part of the book under the general heading of

'The Wear of Rocks by Glaciers', in which he sets out some first-hand observations on abrasion and the origin of rock flour.

Around the turn of the century, the literature contains many important additions to the body of knowledge on glacial geomorphology, whereas, with the exception of the work by J. Tyndall and A. Heim, progress in glaciology during the latter half of the nineteenth century had been slow. In the field of landform studies, W. M. Davis's contributions are particularly significant. Like Forbes, he was a convinced supporter of the hypothesis of glacial erosion. Although his observations were frequently generalised (e.g. the 'glance from a passing train', p. 42) and unsupported by actual measurements, his arguments gained much support and his categorisation and illustration of glacial landforms proved to be the foundation for writings on this subject for the next fifty years. His paper published in the *Proceedings of the Boston Society of Natural History* for July 1900 was the first to set out in detail his arguments for the origin by glacial erosion of such forms as glacial troughs, hanging valleys and cirques, and it is an excellent example of his deductive method. One portion considers the problem of determining how much erosion must be attributed to ice in the production of existing forms, and another expounds on a proposed 'cycle of glacial denudation', speculating (pp. 61–62) on the probable end-product of such a cycle, a point taken up by Linton in 1963 (p. 166).

The other two papers chosen of similar date, one by Willard D. Johnson of 1904 and the other by G. K. Gilbert of 1906, are quite different in their approach – both base theories and deductions on careful field observation and measurement. Johnson's observations on cirques in the Sierra Nevada of California showed the importance of the freeze–thaw process in cirque headwall recession. Such recession had been postulated by Davis and other workers, but the processes responsible remained obscure. Johnson's investigations of processes actively at work in a bergschrund, while they placed too much emphasis on the bergschrund itself, nevertheless were to pave the way for further related investigations by W. V. Lewis in the 1930s and 1940s, W. R. B. Battle in the 1950s, and J. E. Fisher in the 1960s.

Gilbert's paper on 'Crescentic Gouges on Glaciated Surfaces' was concerned with another aspect of glacial erosion, that of the modelling and plucking of hard rock surfaces by moving ice. Considerable attention had been paid to the striation and polishing of such surfaces by T. C. Chamberlin in his great monograph of 1888; Gilbert's was the first analysis of other small-scale forms of bedrock erosion, and has stimulated many other workers to investigate the origins of these interesting forms. Undoubtedly much remains to be learnt about their origins, but most subsequent workers have agreed that Gilbert's theories are still broadly acceptable, a testimony to his careful and objective analysis. In considering the mode of basal ice flow over and around bedrock obstacles, Gilbert's thinking was far in advance of any of his contemporaries and can be linked with much more modern ideas such as are discussed in the papers by Kamb and LaChapelle and Weertman. It is also worth pointing out that Gilbert was perfectly aware (though he could not prove it) that temperatures at the base of a temperate glacier are at pressure melting point. Indeed, on p. 91, he is almost hinting at the theory of regelation slip, to be developed over fifty years later.

From the inter-war years, two articles only have been selected, though once again the reason is not lack of choice. But no anthology on this theme would be complete without reference to the work of François E. Matthes, member of the United States Geological Survey, nor without reference to the area with which this paper deals, Yosemite Valley in California. Indeed, this and the preceding papers by Johnson and Gilbert are all concerned with different aspects of the glacial geomorphology of parts of the Sierra Nevada, but whereas the first two are dominantly concerned with processes and detailed features, Matthes's monograph is the grand study on a regional scale, concerned with the evolution of the whole area from preglacial to postglacial times. The original runs to 160 pages in length and is profusely illustrated; here, it was only possible to pick out a portion, less than one-tenth of the whole, dealing with the extent of modification to Yosemite by glacial erosion. Unlike W. M. Davis, whose writings also include regional studies, Matthes considered in some detail the processes responsible for sculpturing the landscape – note especially his analysis of the 'plucking' or 'quarrying' mechanism of glacial erosion – and attempted, to a far greater extent than Davis, to quantify the amount of erosion achieved by ice. A selection of his careful reconstructions of the most probable preglacial surface (Figs. 5.2–5.8) is included, though it was not possible to include the actual evidence (which occupies several tens of pages in the original monograph) on which these reconstructions are based. One further point in connection with his estimates of the depths to which ice excavated the troughs is that Matthes was not aware (indeed, he could not have been, for seismic exploration techniques were not then available) of the great depths of infilling below the trough floors in places. In Fig. 5.5, Matthes estimates a total depth of vertical glacial excavation of 900 ft. On the line of this section, B. Gutenberg *et al.* (1956) showed that there is 1800 ft of sedimentary infill whereas Matthes only allowed for about 200 ft. His estimate of 900 ft of erosion must therefore now be increased to 2500 ft at this locality, and his estimate of the volume of rock removed from a section of valley 1 yd long from 700,000 to about 1,100,000 yd^3.

The remaining pre-war contribution also deals with the problem of measuring amounts of glacial erosion, but in terms of present-day glaciers over known periods of time. If obtainable, such data might permit useful comparisons to be drawn between rates of erosion attributable to glaciers on the one hand and agents such as rivers on the other. One of the most promising methods of establishing rates of contemporary glacial erosion is by measuring the silt discharge by glacier meltwater streams, for it now seems probable that this silt not only represents largely the product of glacial erosion but also that it represents the bulk of material being produced by erosion in any given glacier basin. The application of the method to Hoffellsjökull, an outlet glacier of Vatnajökull, is described by Sigurdur Thorarinsson, and the practical difficulties involved in the method are well illustrated. Nevertheless, the evidence is sufficiently reliable to suggest that glacial erosion beneath Hoffellsjökull is working at a rate about five times faster than fluvial erosion in an adjacent non-glacierised valley. The paper represents a small part and only one aspect of the valuable results obtained by the Swedish–Icelandic investigations of 1936–8 (H. W:son Ahlmann and S. Thorarinsson), the forerunner of many subsequent important glaciological studies in Iceland.

Of the seven papers selected from the years 1957–65, the first two are from David L. Linton's extensive published work on glacial landforms. The first is in a journal not easily accessible to most students, while the second attempts a broad coverage of the forms of glacial erosion, drawing together the threads of much earlier work. The first of the two papers is reprinted here virtually in its original form. It is concerned to show that, as glacierisation of a landscape progresses from the incipient stage of cirque (corrie) glaciation, through a stage in which valley glaciers are dominant, to a stage in which complete submergence of the relief by ice is attained, so ice movement becomes progressively less constrained by the underlying relief. In the intermediate stages, the outflow from major centres of ice dispersion may succeed in imparting a radiating pattern to the valley system, adapting portions of existing valleys and creating new ones by glacial erosion. Linton shows how this stage may have been reached in the English Lake District, thus offering a hypothesis for the origin of its present-day radial drainage quite different from the classic hypothesis of drainage superimposition from a domed cover.

Linton's paper of 1963 represents his Presidential Address to the Institute of British Geographers. Its length has necessitated some shortening, mainly by omission of sections giving further examples of particular forms. His review concentrates in turn on ice-moulded forms (roches moutonnées, rock drumlins, etc.); glacial troughs and their classification (which provides a link with his 1957 paper on radiating valleys); corries and their progressive enlargement, resulting in gradual destruction of divides; and finally, the ultimate expected results of prolonged glacial erosion where lateral valley divides are overwhelmed by streaming ice and eventually reduced or even eliminated. In this final stage of the reduction of a landscape by glacial erosion, the last remnants of the preglacial relief may survive as pyramidal peaks. Such a stage may well be recognisable in parts of the Antarctic. As already noted, it is instructive to compare Linton's deductions on the results of prolonged glacial erosion with those of W. M. Davis written in 1900 (p. 62 of this volume). Another important theme touched on by Linton is that of the role of dilatation jointing in the evolution of glacial landforms. Produced by unloading of bedrock as erosion progresses, it is argued that these sheet-like joint systems, parallel to the surface of unloading, powerfully affect in turn the direction and rate of erosion. There is evidence that they develop, for instance, parallel with the walls and floors of glacial troughs, so that these are enlarged and deepened by the stripping-off of successive layers of rock and, at the same time, preserve their U-shaped section with little change. Similarly, the outlines of cirques are related to dilatation jointing, and whereas the central portions of arêtes separating adjacent cirques are susceptible to destruction by virtue of the intersection here of different dilatation joint systems, the pyramidal peaks at the junctions of arêtes may be relatively immune owing to their greater distances from the centres of cirque excavation.

A particular instance of an area subjected to intense and prolonged ice erosion in the Pleistocene is the subject of Keith M. Clayton's paper. In the area of the Finger Lakes, New York State, Clayton suggests that the work of ice has been carried to the point where virtually the whole of the present relief is the work of ice and independent of the preglacial form. An important general point is the attempt to recognise zones of different intensity of glacial erosion on the basis of the salient landform characteristics, a concept that has been usefully applied elsewhere. In the Finger Lakes area, three basic zones, A, B and C in increasing order of intensity of erosion, from south to north, are described; zone A is one in which occasional 'through-valleys' (breached preglacial divides) are encountered, while at the other extreme zone C represents an area of intense trenching and destruction of the preglacial relief, where the deep rock basins of the Finger Lakes themselves are to be found. Clayton also argues that, in zones A and B, although the Allegheny Plateau was covered by ice (as shown by the Olean drift), ice erosion was negligible on the plateau in comparison with the valleys trenched in the plateau. This situation represents the phenomenon of 'ice streaming', resulting in strongly selective erosion by the moving ice.

The final group of papers has a more glaciological emphasis, and represents, in my judgement, some of the most significant work of the last decade in its implications for glacial geomorphology. Six authors are involved. J. W. Glen has made major contributions to knowledge in the field of the physics of ice and its deformation under different conditions of stress and temperature. His co-author in the paper describing and analysing the side-slip studies made at Austerdalsbreen in Norway, in 1959, was W. Vaughan Lewis. Lewis worked extensively on glaciers and glacial geomorphology for some twenty-five years, and no anthology concerned with ice and glaciation would be complete without an example from his writings. His death in 1961 was a great loss to both geomorphology and glaciology. Equally tragic was the death of one of Lewis's most able research students, John G. McCall. An engineer by training, he chose for his doctorate research topic a study of the structure and movement of a Norwegian cirque glacier. He approached the problem in the most direct fashion – by organising the excavation of tunnels through the glacier from its surface to its headwall. The editor was one of a band of undergraduates whom John McCall fired with sufficient enthusiasm to hack the main 120 m-long tunnel through the glacier! Barclay Kamb and E. LaChapelle in their paper are also concerned with data derived from a tunnel (through Blue Glacier, Washington), and in particular with observations of the basal sliding process which neatly complement and add to the side-slip observations of Glen and Lewis. The final paper by J. Weertman is included to show the extent to which theoretical concepts in glaciology are keeping up with field observation and experiment, and in turn are capable of throwing useful light on glacial erosion processes.

The papers by Glen and Lewis, McCall, and Kamb and LaChapelle all demonstrate the sorts of problems encountered by field measurements on glaciers. Glen and Lewis describe different ways of measuring side-slip and the differing reliability (and significance) of the results. McCall's paper conveys some impression of the immense amount of work and data collection needed to sustain a thorough investigation of the movement of even a small glacier covering only about 0·1 km^2. The basic problem that all three papers are studying is how a glacier moves over and past an irregular rocky bed; it is reasonable to expect that studies of the mechanisms involved will in turn help to elucidate the processes by which moving ice can erode its bed and channel sides. All three glaciers chosen for study are actively slipping over their beds; basal sliding accounts for 90 per cent of total movement in the cases of Vesl-Skautbreen (studied by McCall) and Blue Glacier, while side-slip at Austerdalsbreen ranges from 10–65 per cent of maximum centre-line ice velocities (compare Forbes's findings, pp. 31–2, which gave a maximum of 70 per cent, though he had no marker closer to the ice edge than 100 yd). Glen and Lewis's findings support Forbes also in that a correlation seemed to exist between rates of side-slip and weather conditions, suggesting that meltwater finding its way down the side wall and to the bed of the glacier had a most important role as a lubricant. The final section of Weertman's theoretical discussion takes up this point, and it seems likely that quite small increases in the thickness of a basal meltwater film might cause disproportionately greater increases in rates of glacier sliding. This, of course, only applies to temperate glaciers, and it is important to stress that, because of the paucity of data on cold glaciers, no paper could be included in this anthology to exemplify the very different conditions relating to cold glaciers.

The form of the bedrock surface over which a glacier is sliding has a profound influence on the rate of ice movement, as Weertman's discussion of the 'controlling obstacles' (p. 250) shows. But it is also possible, in certain cases at least, for the type of ice movement to play a part in determining the shape of the eroded bedrock surface.

McCall's studies showed conclusively that the ice of Vesl-Skautbreen was moving, as a whole, in a rotational slip, as Lewis and others had suspected many years previously. McCall, and J. M. Clark and Lewis (1951), considered that such rotational movements might well be instrumental in scouring out basins in the rock floor beneath and would certainly contribute to the transport of debris up and over the cirque lip. However, McCall found no evidence of the existence of shear planes of differential rotational movement within the ice, as had been earlier postulated.

The micro-forms of the ice–rock contact merit attention for the information they may provide on possible erosion processes. It is well recognised that glacial erosion consists of two distinct sets of processes, one termed abrasion, the other plucking, quarrying or joint-block removal. It seems likely, though there is little actual evidence, that the second group is quantitatively the more important, especially in rocks which are neither too coarsely nor too finely jointed, as Matthes observed (p. 107). McCall considered that the form of the bedrock surface exposed at the inner end of the lower tunnel through Vesl-Skautbreen was more suggestive of abrasion than plucking at present, and the instrument of abrasion was clearly the 30 cm-thick debris-laden 'sole' of the glacier (cf. Forbes's observations, p. 36). Kamb and LaChapelle describe how the structure and texture of the lowest layer of ice (up to about 3 cm thick) next to the bedrock beneath Blue Glacier gave evidence of repeated freezing and thawing or 'regelation', and considered that regelation slip was a major mechanism by which ice succeeded in moving over irregular bedrock. On the upstream side of obstacles, the greater pressure induced melting, whereas on the lee side refreezing might occur, especially in the commonly found subglacial cavities on the lee side. Meltwater would flow as a thin film around the obstacle in the direction of glacier motion, while heat would move through the obstacle in the reverse direction. Thus, as Weertman shows in his computations, the larger the obstacle the less efficient the mechanism of regelation slip becomes. On the other hand, for smaller obstacles in jointed rock, the freeze–thaw involved in the regelation mechanism may also play a part in loosening particles of bedrock for transport by the ice.

Glen, Lewis and McCall also consider the stress situation set up when a boulder carried in the basal layers of a glacier encounters a bedrock obstacle. The larger the boulder, the greater the force that the ice can exert on it. Because of the physical nature of ice, there is a limit to the amount of pressure that it can exert on a given area – beyond that limit, the ice 'yields' at an increasingly rapid rate. Glen, Lewis and McCall suggest that a thickness of about 22 m of ice will generate very nearly the maximum possible pressure on any boulder or obstacle – greater thicknesses will have little greater effect. But the force transmitted from a moving boulder to a bedrock obstacle, so long as the ice is at least 22 m thick, will depend primarily on the size of the boulder – i.e. the bearing surface that it presents to the ice. If the boulder is large enough, the yield strength of the obstacle may be exceeded and the obstacle will be sheared off. Alternatively, the boulder may be arrested by the obstacle, when the ice will flow around both. In time, more boulders may be brought along and wedged around the obstacle. Then either the accumulation will grow in size and be streamlined to form, for example, a drumlin by the flow of ice around it, or the combined bearing surfaces of all the newly added boulders will generate sufficient stress for the obstacle to be at last planed off. McCall's calculations of the possible stresses involved are interesting and probably realistic. In the process, of course, rock jointing will be highly significant, as Glen and Lewis emphasise; in particular, the dilatation joints discussed by Linton (p. 162) and Lewis (1954) will greatly help to weaken the bedrock for this form of joint-block removal.

Another section of McCall's paper deals with cirque headwall sapping. Space precludes a complete discussion here, but the student should carefully compare McCall's conclusions with those of W. D. Johnson over fifty years previously (p. 75), and note that McCall considered headwall sapping by freeze–thaw action, in the case of Vesl-Skautbreen, to be quantitatively much more effective at the present day than basal corrasion by grinding or 'plucking'.

Kamb and LaChapelle distinguish, both in laboratory experiments and in the field, between two fundamental mechanisms of basal sliding. The first of these, regelation slip, has already been mentioned; the second derives from the ability of ice to deform under stress, to exhibit behaviour known as 'creep'. Creep, which depends on the physical properties of ice, was first investigated in detail by Glen (1955) who derived a power flow law to express its deformation, in which the strain rate is proportional to the *n*th power of the applied stress. The values of *n* for glacier ice are usually between 2 and 4. This means that, with any increase in the stress applied, the rate of creep will rise comparatively rapidly. Hence, at the base of a glacier, the creep mechanism allows ice to move over and around large obstacles whose size is too great for regelation slip to be efficient. Furthermore, beneath cold glaciers, creep is the only mechanism of basal sliding available.

The theory of this double mechanism for the basal sliding of temperate glaciers is investigated in the final paper by Weertman. Where bedrock obstacles are of various sizes, the speed of sliding will be determined, in effect, by those obstacles whose size permits the rate of flow by regelation slip to equal the rate of flow by creep. These are termed the 'controlling obstacle' sizes. For smaller obstacles, the ice will move over them mainly by regelation slip; for larger ones creep will dominate. The total speed of sliding, although mainly determined by the controlling obstacles, will, however, also be affected by larger obstacles to some extent. Allowing for this, Weertman derives the relationship (8b), p. 253. Figs. 13.1 and 13.2 demonstrate the relationships graphically. It is of great interest that, using data on bed roughness and basal shear stress obtained by Kamb and LaChapelle in the tunnel beneath Blue Glacier, the speed of sliding predicted by Fig. 13.2 is 2 m a year and the predicted size of the controlling obstacles is 4 cm. Kamb and LaChapelle, in comparison, *measured* a speed of sliding of 5·8 m a year, and a regelation layer up to 3 cm thick. The predicted and observed values are at least of the same order of magnitude.

As I pointed out earlier, even the most recent papers selected in this volume do not necessarily represent current opinion, which at a time of great advance in any science is perforce changing rapidly. Since Weertman wrote in 1964, the *Journal of Glaciology* has carried further discussions and argument on the part of Weertman, Lliboutry and others concerning the mechanics and theory of glacier sliding, which it would be impossible to include in this anthology. But to understand these controversies, study of Weertman's 1964 paper is an essential prerequisite. It is also evident that a better understanding of how glaciers slide over irregular rocky beds is most likely to lead to improvements in our theories of glacial erosion, and that theoretical studies (such as those of Weertman), laboratory studies (such as those outlined by Kamb and LaChapelle), and field studies (such as Glen and Lewis's experiments) must go hand in hand. In the field of landform studies, further advances seem likely to come from more quantitative analyses of the forms themselves; whereas morphometric studies in fluvial geomorphology have been developing steadily over the last twenty-five years, the morphometry of glacial landforms has been relatively neglected. In this field, as in the field of process studies, the major advances of the next decade are likely to lie.

1 Account of a Survey of the Mer de Glace and its Environs

JAMES D. FORBES

IT was the especial object of my journey in 1842 to observe accurately the rate of motion of some extensive glacier at different points of its length and breadth.

We have seen that the *motion* of glaciers has been for much more than half a century universally admitted as a physical fact. It is, therefore, most unaccountable that the *quantity* of this motion has in hardly any case been even approximately determined. I rather think that the whole of De Saussure's writings contain no one estimate of the annual progress of a glacier, and if we refer to other authors we obtain numbers which, from their variety and inaccuracy, throw little light on the question. G. J. Hugi (1842) perceived the errors arising from a confusion between the rate of *apparent* advance of an increasing glacier into a warm valley, whilst it is continually being shortened by melting, and the rate of motion of the ice itself. He points out the correct method of observation; and although his work contains no accurate measures, he was perhaps the first who, by observing the position of a remarkable block upon the glacier of the Aar, indicated how such observations might be usefully made, instead of trusting (as appears to have been the former practice) to the vague reports of the peasantry. Hugi's observations on the glacier of the Aar give a motion of 2200 ft in nine years, or about 240 ft/annum. Now, in contradiction to this, it would appear from M. Agassiz's observations (1840), that from 1836–9, it moved as far as in the preceding nine years – that is, three times as fast. There is reason, however, to think, that M. Hugi's estimate is the more correct.

I had myself been witness to the position, in 1841, of the stone whose place had been noted by Hugi fourteen years before, and it was manifest that it had moved several thousand feet. In conformity with the prevalent view of the motion of the ice being perceptible chiefly in summer, I made the hypothesis that the annual motion may be imagined to take place wholly during four months of the year with its *maximum* intensity, and to stand still for the remainder. With this rude guide, and supposing the annual motion of some glaciers to approach 400 ft a year (as a moderate estimate from the previous data), we might expect a motion of at least 3 ft a day for a short time in the height of summer. There appeared no reason why a quantity ten times less should not be accurately measured, and I, therefore, felt confident that the laws of motion of the ice of any glacier in its various parts, and at different seasons, might be determined from a moderate number of daily observations.

I went to Switzerland, therefore, fully prepared, and not a little anxious to make an experiment which seemed so fruitful in results, and though so obvious, still unattempted.

The unusually warm spring of 1842, gave me hopes of commencing my operations earlier than the glaciers are usually frequented; and it was evident, that, to detect the effect of the *seasons* on the motion of the ice, they could not be too soon begun. I left Paris on the 9th of June, by the *malle poste* for Besançon. After spending a day at Neufchâtel, I proceeded to Berne to visit M. Studer, and from thence I went to Bex, to make the acquaintance of M. de Charpentier, with whose geological and other writings I had so long been familiar. I only allowed myself a hasty visit to my friends at Geneva, and left that town with lowering weather, on the 23rd June, for Chamouni, determined to await its clearing, and then proceed at once to the Mer de Glace. No patience was, however, required. The weather cleared that very day, and reaching Chamouni early on the following one, I made the requisite arrangements at the village, and leaving my baggage to follow, I proceeded straight to the Montanvert.

I resolved to commence my experiments with the very simple and obvious one of selecting some point on the surface of the ice, and *determining its position with respect to three fixed co-ordinates*, having reference to the fixed objects around; and, by the variation of these, to judge of the feasibility of the plans which I had laid out for the summer campaign. One day (the 25th) was devoted to a general reconnaissance of the glacier, throughout a good part of its length, with a view to fixing permanent stations; and the next I proceeded, at an early hour, to the glacier opposite to the rocky promontory on the west side of the Mer de Glace, called *L'Angle*, thirty minutes' walk from the Montanvert, which presented a solid wall of rock in contact with the ice, so that upon the former, as upon a fixed wall or dial, might bc marked the progress of the glacier as it slid by. The instrument destined for the observations was the small astronomical circle, or 4½ in. theodolite, supported on a portable tripod. A point of the ice whose motion was to be observed, was fixed by a hole pierced by means of a common blasting iron or *jumper*, to the depth of about 2 ft. At first, I was much afraid of the loss of the hole by the melting of the ice, and the percolation of water from day to day; but I soon found that very little precaution was necessary on this account, and that such a hole is really a far more permanent mark than a block of stone several tons in weight resting on the ice, which is very liable to change of position, by being raised on a pedestal, and finally slid into some crevasse.

An accurate vertical hole being made, the theodolite was nicely centred upon it by means of a plumb line, and levelled. A level run directly to the vertical face of rock, gave at once the co-ordinate for the *vertical* direction, or the height of the surface of the glacier. The next element was the position or co-ordinate parallel to the length or direction of motion of the glacier. This was obtained by directing the telescope upon a distant fixed object, nearly in the direction of the declivity of the glacier, and which object was nothing else than the south-east angle of the house at the Montanvert, distant 5000 ft. The telescope was then moved in azimuth exactly 100° to the left, and thus pointed against the rocky wall of the glacier, which was here very smooth and nearly perpendicular, owing to the friction of the ice and stones. My assistant was stationed there with a piece of white paper, with its edge vertical, which I directed him by signs to move along the surface of the rock until it coincided with the vertical wire of the telescope. Its position was then marked on the stone with a pencil, and the positions of successive pencil marks were carefully measured by a tape or ruler from day to day. Marks were then indented in the rock with a chisel, and the mark painted red with oil paint, and the date affixed. These marks, it is believed, will remain for several years. The station on the ice was distant 250 ft from the rock, and, by repeating the observation frequently, I found that it could be depended on to about one-fourth or one-third of an inch.

The third co-ordinate, or that which should measure the distance of the station from the rock was not so accurately ascertained. No ready means offered itself for ascertaining with quickness and accuracy any variation of distance in respect to the breadth of the glacier. Whilst I admit that this would have been an advantage, I may observe that in most cases there is no reason to doubt that the motion of the ice is sensibly parallel to its length, and that any small error in the direction would scarcely affect the result. The direction of motion of the ice is unequivocally proved by the direction of the moraines, which are an external indication of that motion. In general, therefore, I have measured the movement of the ice parallel to the moraines where they were well marked. I am of opinion, however, that a check of some kind, such as the measurement of a third co-ordinate, would be advantageous where applicable.

THE MOTION OF THE MER DE GLACE

It was with no small curiosity that I returned to the station of the 'Angle' on the day following the first observation. The instrument being pointed, and adjusted as already described, and stationed above the hole pierced in the ice the day before, when the telescope was turned upon the rock the red mark was left far above, the new position of the glacier was 16·5 in. lower (that is, more in advance) than it had been 26 hr previously. Though the result could not be called unexpected, it filled me with the most lively pleasure. The diurnal motion of a glacier was determined (as I believe) for the first time, from observation, and the methods employed left no doubt of its being most accurately determined. But a question of still greater interest remained behind. Was this motion a mean and continuous one, or the result of some sudden jerk of the whole glacier, or even the partial dislocation of the mass of ice on which I stood? This could only be tested by successive days' trial, and I awaited the result with doubt and curiosity. Of this I was persuaded, that if the motion should appear to be continuous, and *nearly* uniform, it could not be due to the mere sliding of the entire glacier on its bed, as De Saussure supposed; for, admitting the possibility of gravity to overcome such intense friction as the bed of a glacier presents, it seemed to me quite inconsistent with all mechanical experience that such a motion, unless so rapid as to be an accelerated one, and that the glacier should slide before our eyes out of its hollow bed (which would be an avalanche), could take place, except discontinuously, and by fits and starts. To this most elementary question no answer founded on direct experience is to be found, so far as I know, in any work; and although the whole theory might turn upon so simple a point, as whether the glacier *flows* down evenly, or moves by jerks, opinions seem hitherto to have been divided. On the 28th June 1842 I therefore hastened with not less interest to my post, and found that in 25½ hr the advance had been 17·4 in., nearly the same, though somewhat more rapid, than on the previous day. I no longer doubted that the motion was continuous, but I hastened to put it to a still more severe test. I proposed to compare the *diurnal* and *nocturnal* march. I fixed its position at 6 p.m. on the 28th, and next morning by 6 o'clock I was again stationed on the glacier. It had moved 8 in., or exactly half the mean daily motion already observed. The night had been cold; the ice was still frozen, though the temperature of the air had already risen to 40°F; a thermometer laid on the ice stood at 36°F. If congelation had resulted, during the night, so as to freeze the water in the capillary fissures, nearly the whole motion of the 24 hr ought to have taken place whilst the glacier froze: but not at all: from 6 a.m. to 6 p.m. of the 28th, the glacier advanced 9·5 in., giving a total motion of 17·5 in. in 24 hr, somewhat greater than either of the preceding days, the motion appearing to increase as the warm weather continued and increased in intensity: at least so I interpreted it. The same afternoon I had no difficulty in detecting the advance of the glacier, during an interval of *an hour and a half*. The continuity of motion was thus placed beyond a doubt. The marks on the rock indicated a regular descent in which time was marked out as by a shadow on a dial.

The following morning (30th June) at 6 o'clock, the glacier was 8·5 in. in advance, and during the succeeding 12 hr of day, 8·9 in., making together 17·4 in. for the 24 hr, a result not differing sensibly from that of the day before.

I observed distinctly the progress of the glacier on the 30th from 5–6 p.m., and on this occasion, as on the day before, it appeared to me that the motion at that time of day was *more rapid* than the mean motion. The motion in 24 hr for these 4 days had been:

15·2, 16·3, 17·5, 17·4 in.,

a variation which I believed to be by no means accidental, but due to the increasing heat of the weather.

These results were the more interesting (and with respect to their regularity the more unexpected) because the spot where they were made was a part of the ice deeply crevassed. It had been selected on account of the proximity of the naked rock; but though the most solid accessible part of the ice was chosen for the station A, it was surrounded by chasms in every direction, and the glacier in nearly all its breadth between the Angle and the Echellets is (in ordinary language) impassable on account of its dislocated and shattered condition. Yet amidst all this turmoil and confusion there were no fits of advance, no halts, but an orderly continuous progression.

But during the last week of June, in which, stimulated by the extraordinary fineness of the weather, and the fresh interest of every day's experiments, I spent from 12–14 hr daily on the glacier – I was able to make other observations of interest to the theory, and not less consistent with one another. I fixed two points in the ice by bored holes a little way below the Montanvert, one near the side, the other near the centre of the glacier. Most authors, I believe, have asserted, that the *sides* of the glacier move faster than the *centre*. (See, for example, L. Agassiz (1840) p. 167.) But this seemed worthy of proof. Stationing my theodolite, not upon the ice, but upon the lofty western bank at the station D, on a great boulder 60 yd in a direction north, 40° east (magnetic) from the south-east corner of the house of the Montanvert, I levelled it carefully, and then turning the telescope so as to point *across* the glacier to the rocks on the opposite side, by unclamping the telescope I caused it to describe a vertical great circle. I caused a tall cross (D 1) to be painted in red bordered with white on a face of rock opposite, making an angle of 118° with the corner of the Montanvert already mentioned, and distant from D 2898 ft.

By pointing the telescope upon the cross, and then causing it to describe a vertical circle (like a transit instrument adjusted upon a meridian mark) the velocity of the different parts of the glacier could be determined as they flowed past. Two stations, as has been said, were first fixed upon and marked by vertical holes in the ice renewed from time to time; the first D 2 was about 300 ft from the west bank of the glacier, therefore, nearly corresponding in position to station A, which was 5200 ft higher up; the other, marked D 3, was 795 ft farther east, or rather beyond the centre of the glacier, being within 150 ft of the first moraine. It is, however, very near the centre.

	Side (D 2)	Centre (D 3)
From 29th June to 1st July the motion in 24 hr, was	17·5 in.	27·1 in.

Here, then, was a difference not to be mistaken, and the near coincidence of the side station with the result at station A, I considered at the time confirmatory of its accuracy. Henceforth, I entertained no doubt that the generally received opinion is incorrect, and that the glacier stream, like a river, moves fastest towards its centre.

In the same line across the glacier with D 2 and D 3, several other stations were afterwards fixed with a view to test the modification of velocity depending on the distance from the bank or edge of the glacier. These measures proved that the velocity of the central parts is nearly alike, and that the greatest differences in velocity are close to

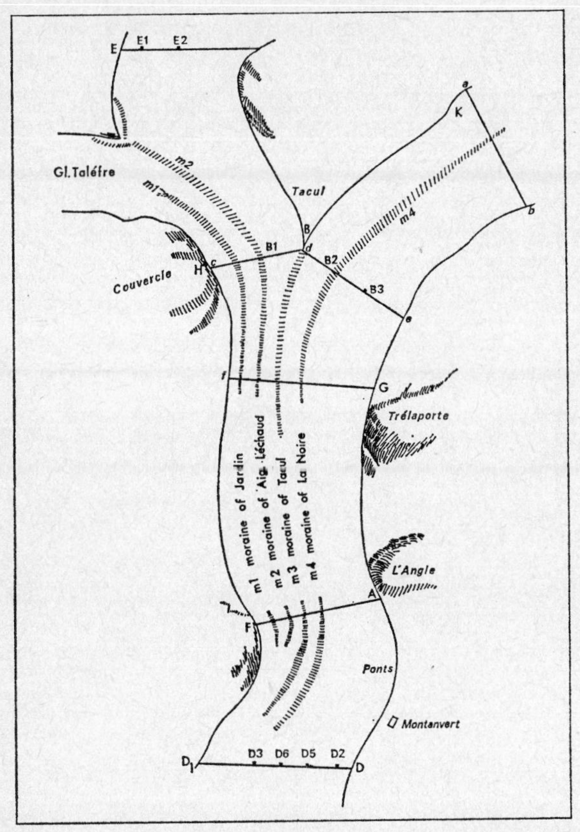

Fig. 1.1

the side, where friction may be expected to act exactly as in a current of water.

At later periods, there were added further points of observation (shown on Fig. 1.1).

A careful examination of Table 1.2 will confirm the following deductions, more full and explicit than those which my first week's observations afforded, and which lay down, I believe for the first time, the General Laws of the Motion of a Glacier deduced from observation.

Table 1.1 Glacier motion

(Reckoned in each case from the commencement of the observation)

Near Montanvert, (1) D 2		Near Montanvert, (2) D 4		Near Montanvert, (3) D 6		Near Montanvert, (4) D 3		L'Angle A	
1842	ft	1842	ft	1842	ft	1842	ft	1842	ft
June 20	0	July 28	0	Sept 17	0	June 29	0	June 26	0
July 1	2·9	Aug 1	7·0	20	4·9	July 1	4·5	27	1·2
28	43·2	9	23·2	26	14·9	28	64·5	28	2·6
Aug 1	48·8			28	18·9	Aug 1	71·7	29	4·1
9	59·6					Sept 16	163·6	30	5·5
Sept 16	116·6	D5				17	165·5	July 28	38·4
17	118·0					20	170·5	Aug 1	42·9
18	119·2	Sept 17	0			26	180·7	9	52·7
19	120·3	19	3·1			28	184·6	Sept 16	9·40
20	121·6	20	4·8					26	10·32
26	128·6	26	14·4						
28	132·0	28	18·6						

Table 1.1 (*contd.*)

Glacier de Léchaud, Pierre Platte C		Glacier de Léchaud B 1		Glacier du Géant B 2		Glacier du Géant B 3		Glacier du Léchaud E 1		Glacier de Léchaud E 2	
1842	ft	1842	ft	1842	ft	1842	ft	1842	ft	1842	ft
June 27	0	June 30	0	June 30	0	Aug 4	0	July 29	0	July 29	0
30	2·5	Aug 2	29·6	Aug 2	37·8	6	3	Aug 2	3·7	Aug 2	4·0
Aug 2	29·9	6	32·9	4	40·2	Sept 17	47	8	10·9	8	12·0
Sept 17	63·2	Sept 17	66·9	6	42·5			Sept 25	56·0		
				Sept 17	79·1						

Table 1.2 Mean daily glacier motion

D 2

		in.
June	29– 1 July	17·5
July	1–28	17·3
	28– 1 Aug	16·2
Aug	1– 9	16·6
	9–16 Sept	18·0
Sept	16–17	16·9
	17–18	13·8
	18–19	13·1
	19–20	16·3
	20–26	14·2
	26–28	19·5

D 4

		in.
July	28– 1 Aug	21·0
Aug	1– 9	24·7

D 5

		in.
Sept	17–19 Sept	18·6
	19–20	20·3
	20–26	19·2
	26–28	25·2

D 6

		in.
Sept	17–20 Sept	19·7
	20–26	20·1
	26–28	23·7

D 3

		in.
June	29– 1 July	27·1
July	1–28 July	25·7
	28– 1 Aug	21·0
Aug	1–16 Sept	24·0
Sept	16–17	23·7
	17–20	20·3
	20–26	20·4
	26–28	22·5

A

		in.
June	26–27 June	15·2
	27–28	16·3
	28–29	17·5
	29–30	17·4
	30–28 July	14·0
July	28– 1 Aug	13·6
Aug	1– 9	15·4
	9–16 Sept	13·0
Sept	16–26	11·15

C

		in.
June	27–30 June	10·2
	30– 2 Aug	9·9
Aug	2–17 Sept	8·7

B 1

		in.
June	30– 2 Aug	10·8
Aug	2– 6	10·0
	6–17 Sept	9·7

B 2

		in.
June	30– 2 Aug	13·8
Aug	2– 4	14·0
	4– 6	14·25
	6–17 Sept	10·4

B 3

		in.
Aug	4– 6 Aug	18·0
	6–17 Sept	12·6

E 1

		in.
July	29– 2 Aug	11·3
Aug	2– 8	14·3
	8–25 Sept	11·3

E 2

		in.
July	29– 2 Aug	13·5
Aug	2– 8	16·3

THE LAWS OF GLACIER MOTION

I. *The motion of the higher parts of the Mer de Glace is, on the whole,* slower *than that of its lower portion; but the motion of the middle region is slower than either.*

I had not failed to point out, when I proposed the determination of the velocity of different points of a glacier, as a test of the cause of its motion, that this must depend materially upon the form of its section at different parts. The velocity of a river is greatest where it narrows, and is small in the large pools. Just so in the Mer de Glace. It is truly a vast magazine of ice, with a comparatively narrow outlet, as the map distinctly shows; the two glaciers of the Géant and Léchaud, uniting just above the strait formed by the promontories of Trélaporte and the Couvercle. Hence results, as we have seen, the great ice basin, where we have reason to conclude (as before observed), that the glacier attains a greater thickness than at any other part, and thus, though the breadth of the two confluent glaciers taken separately is greater than after their union, being, undoubtedly, much shallower there, their area of section is smaller, and therefore the velocity of the ice will be greater. There will, indeed, be always a *condensation* of the ice within the triangle *BHG*, owing to the resistance opposed to its egress; and here, accordingly, the surface of the ice is most level. It is not indeed strictly true, that the quantity of ice passing through any section of the glacier in a given time, is exactly equal; because there is fusion and evaporation, amounting to an actual loss of substance, between any two sections, and this becomes especially obvious near the lower extremity of the glacier. There is, therefore, no ground for surprise at the fact, that the middle part of the glacier moves forward slower than the higher parts. Had the glacier *continued* to expand in breadth, as very many glaciers do, no check would have occurred, and the anomaly would have disappeared.

II. *The Glacier du Géant moves faster than the Glacier de Léchaud, in the proportion of about seven to six.* The vast mass of the former glacier tends to overpower the other, in some measure, and it takes the lion's share of the exit through the strait between Trélaporte and the Couvercle, squeezing the ice of Léchaud and Taléfre united, into little more than one-third of the breadth of the whole. It is to this circumstance that I impute the excessively crevassed state of the eastern side of all the Mer de Glace, which renders it almost impossible to be traversed; the ice is tumultuously borne along, and, at the same time, squeezed laterally by the greater velocity and mass of the western branch.

III. *The centre of the glacier moves faster* (as we have seen) *than the sides.* When two glaciers unite, they act as a single one in this respect, just as two united rivers would do. Now this variation is most rapid near the sides, and a great part of the central portion of the glacier moves with no great variation of velocity. Thus we find that four stations taken in order, from the side to the centre of the glacier (or a little beyond it), have the following rates of motion:

Station	D 2	D 4/5	D 6	D 3
	1·000	1·375	1·356	1·398

Or if we compare observations made all at the same season of the year (September), we shall find the increase of velocity in every case,

Station	D 2	D 4/5	D 6	D 3
	1·000	1·332	1·356	1·367

The first point was 100 yd from the edge of the glacier; the next 130 yd farther. In this short space the velocity had increased above a third part.

The explanation which we offer of this, as due to the friction of the walls of the glacier, would lead us to expect such a law of motion. The retardation of a river is chiefly confined to its sides; the motion in the centre is comparatively uniform.

Similar reasoning would lead us to expect that (supposing the glacier to slide along its base) the portions of ice in contact with the bed of the valley will be retarded, and the superficial parts ought to advance more rapidly. The change of velocity in this case also, will be greatest near the bottom.

IV. *The difference of motion of the centre and sides of the glacier varies* (1) *with the season of the year, and* (2) *at different parts of the length of the glacier.*

(1) The following numbers show the velocity ratios of the centre and side of the glacier, near the Montanvert, at the marks D 3 and D 2, during different parts of the season 1842:

	Relative velocity, D 3:D 2
June 29–July 1	1·548
July 1–July 28	1·489
July 28–September 16	1·349
September 16–September 28	1·367

In general, therefore, *the variation of velocity diminished as the season advanced*; we shall presently show that it was very nearly proportional to the *absolute* velocity of the glacier at the same time.

(2) *The variation of velocity with the breadth of the glacier* is least considerable in the higher parts of the glacier or near its origin. Thus, if we compare the velocities of station C, and the mark B 1 on the Glacier de Léchaud near the Tacul, the former being near the side, the latter near the centre of the glacier, we find

	Relative velocity, B 1:C
June 30–August 2	1·09
August 2–September 17	1·12

Again, higher up the same glacier, opposite E, we have the velocity ratios at the centre and side of the glacier:

	E 2:E 1
July 29–August 2	1·19
August 2–August 8	1·14

This ratio is indeed a little greater than the preceding, which corresponds with the fact which we have already found, that the absolute velocity of the glacier is greater at E than at C. Hence, it is highly probable in every case that *the variation of velocity in the breadth of a glacier is proportional to the absolute velocity, at the time, of the ice under experiment*. This is further confirmed by the velocities of the Glacier du Géant at the marks B 2 and B 3, of which the former is near the side and the latter near the centre:

	Velocity ratio, B 3:B 2
August 4–August 6	1·30
August 6–September 17	1·21

Now the absolute velocity of this glacier is greater than that of Léchaud, but less than that at the Montanvert.

V. *The motion of the glacier generally, varies with the season of the year and the state of the thermometer*. Perhaps the most critical consideration of any for the various theories of glacier motion is the influence of external temperature upon the velocity. In this respect my observations, though confined only to the summer and autumn, are capable of giving pretty definite information. Indeed, one circumstance which on other accounts I had much reason to regret, I mean the rigorous weather of the month of September, which hindered many of my undertakings, gave me an opportunity of observing the effect of the first frosts, and thus establishing some important facts as to the influence of cold and wet upon the glacier. This I apprehend to be clearly made out from my experiments, *that thawing weather and a wet state of the ice conduces to its advancement, and that cold, whether sudden or prolonged, checks its progress*. The rapid movement in the end of June which is perceptible at D 2, D 3, A and C, is due to the very hot weather which then occurred, and the very marked reduction at the end of July, to a cold week which occurred at that period. The striking variations in September, especially at the lower stations, which were frequently observed, prove the connection of temperature with velocity to demonstration.

During the continuance of the cold weather, accompanied by snow, from the 18th to the 27th September, it will be observed that the glacier motion was visibly retarded at all the lower stations which were then observed. During this period the thermometer fell at the Montanvert to 20°F; but when mild weather set in again, the glacier became clear of snow (which took place in the lower part on the 27th) and being thoroughly saturated with moisture, it resumed a march as rapid as that of the height of summer.

A THEORY OF GLACIER MOTION

After the detailed though scattered deductions which have been made in the course of this work, from observations on the Movement and Structure of Glaciers, as to the cause of these phenomena, little remains to be done but to gather together the fragments of a theory for which I have endeavoured gradually to prepare the reader, and by stating it in a somewhat more connected and precise form, whilst I shall no doubt make its incompleteness more apparent, I may also hope that the candid reader will find a general consistency in the whole, which, if it does not command his unhesitating assent to the theory proposed, may induce him to consider it as not unworthy of being further entertained.

My theory of Glacier Motion then is this: *A Glacier is an imperfect fluid, or a viscous body, which is urged down slopes of a certain inclination by the mutual pressure of its parts.*

The sort of consistency to which we refer may be illustrated by that of moderately thick mortar, or of the contents of a tar-barrel, poured into a sloping channel. Either of these substances, without actually assuming a level surface, will *tend* to do so. They will descend with different degrees of velocity, depending on the *pressure* to which they are respectively subjected – the *friction* occasioned by the nature of the channel or surface over which they move – and the *viscosity*, or mutual adhesiveness, of the particles of the semifluid, which prevents each from taking its own course, but subjects all to a mutual constraint. To determine completely the motion of such a semifluid is a most arduous, or rather, in our present state of knowledge, an impracticable investigation. Instead, therefore, of aiming at a cumbrous mathematical precision, where the first data required for calculation are themselves unknown with any kind of numerical exactness, I shall endeavour to keep generally in view such plain mechanical principles as are, for the most part, sufficient to enable us to judge of the comparability of the facts of Glacier Motion with the conditions of viscous or semifluid substances.

Now, in all these respects, we have an exact analogy with the *facts* of motion of a glacier, as observed on the Mer de Glace.

First, we have seen that the centre of the glacier moves faster than the sides (p. 32). We have not indeed extended the proof to the top and bottom of the ice-stream, for it seems difficult to make this experiment in a satisfactory manner. In the case of a glacier 600 ft deep, the upper 100 ft will move nearly uniformly, on the principles already mentioned; hence, crevasses, formed from year to year, will not incline sensibly forwards on this account, especially as the action of trickling water is to maintain the verticality of the sides. I have no doubt that glaciers slide over their beds, as well as that the particles of ice rub over one another, and change their mutual positions.

Secondly, the chief variation of velocity is, we have seen, near the sides.

Thirdly, the amount of lateral retardation depends upon the actual velocity of the stream under experiment; whether we consider different points of the glacier, or the same point at different times.

Fourthly, the glacier, we have seen, like a stream, has its still pools and its rapids. Where it is embayed by rocks, it accumulates – *its declivity diminishes, and its velocity at the same time*; when it passes down a steep, or issues by a narrow outlet, its velocity increases.

The *central* velocities of the lower, middle, and higher regions of the Mer de Glace are:

1·398 0·574 0·925

And if we divide the length of the glacier into three parts, we shall find something like these numbers for its declivity:

15° 4½° 8°

Lastly, when the semifluid ice inclines to solidity during a frost, its motion is checked; if its fluidity is increased by a thaw, the motion is instantly accelerated. Its motion is greater in summer than in winter, because the fluidity is more complete at the former than at the latter time. The motion does not cease in winter, because the winter's cold penetrates the ice as it does the ground, only to a limited extent. It is greater in hot weather than in cold, because the sun's heat affords water to saturate the crevices: but the proportion of velocity does not follow the proportion of heat, because any cause, such as the melting of a coating of snow by a sudden thaw, as in the end of September 1842, produces the same effect as great heat would do. Also, whatever cause accelerates the movement of the centre of the ice increases the difference of central and lateral motion.

THE WEAR OF ROCKS BY GLACIERS

There can be no doubt from observation, that a glacier carries along with its inferior surface a mass of pulverised gravel and slime, which, pressed by an enormous superincumbent weight of ice, *must* grind and smooth the surface of its rocky bed. The peculiar character of glacier water is itself a testimony to this fact. Its turbid appearance, constantly the same from year to year, and from age to age, is due to the impalpably fine *flour* of rocks ground in this ponderous mill betwixt rock and ice. It is so fine as to be scarcely depositable. No one who drives from Avignon to Vaucluse can fail to be struck with the contrast of the streams, artificially conveyed on one and on the other side of the road, in order to irrigate the parched plain of Provence. The one is the incomparably limpid water of Petrarch's fountain; the other an offset from the river Durance, which has carried into the heart of this sun-burnt region the unequivocal mark of its birth amidst the perpetual snows of Monte Viso. This is the pulverizing action of ice.

Most erroneously have those argued who object to this theory that ice cannot scratch quartz – ice is only the *setting* of the harder fragments, which first round, then furrow, afterwards polish, and finally scratch the surface over which it moves. It is not the wheel of a lapidary which slits a pebble, but the emery with which it is primed. The gravel, sand and impalpable mud are the emery of the glacier.

When the ice of the Brenva glacier abuts against the foot of Mont Chétif, it is violently forced forward, as if it would make its way up the face of the hill. Here the contact of the ice and soil is very well seen; and we were able to discover a point of contact between the limestone and a protuberant mass of ice which admitted of easy removal, thus showing the immediate action of the ice and rock. The soil near the ice appeared to have been but recently exposed by the summer's melting of the ice. It was chiefly composed of clayey debris from the blue limestone. A piece of fixed rock opposed the ice, and was still partly covered by a protuberance of the glacier, which we speedily but gently cut away with the hatchet. The ice removed, a layer of fine mud covered the rock, not composed, however, alone of the clayey limestone mud, but of sharp sand, derived from the granitic moraines of the glacier, and brought down with it from the opposite side of the valley. Upon examining the face of the ice removed from contact with the rock, we found it *set* all over with sharp angular fragments, from the size of grains of sand to that of a cherry, or larger, of the same species of rock, and which were so firmly fixed in the ice as to demonstrate the impossibility of such a surface being forcibly urged forward without sawing and tearing any comparatively soft body which might be below it. Accordingly, it was not difficult to discover in the limestone the very grooves and scratches which were in the act of being made at the time by the pressure of the ice and its contained fragments of stone. By washing the surface of the limestone we found it delicately smoothed, and at the same time furrowed in the direction in which the glacier was moving, that is, against the slope of the hill.

2 Glacial Erosion in France, Switzerland and Norway

WILLIAM MORRIS DAVIS

EIGHTEEN years ago I presented to this Society an essay on Glacial Erosion, in which my own observations were supplemented by a review of all that I could find written on the subject in the hope of reaching some safe conclusion regarding what was then (as it is still) a mooted question. Although recognizing effective erosion to depths of 'a moderate number of feet' where ice pressure was great and motion was rapid, in contrast to deposition where pressure and motion were reduced and where the amount of subglacial drift was excessive, I could not at that time find evidence to warrant the acceptance of great glacial erosion, such as was advocated by those who ascribed Alpine lakes and Norwegian fiords to this agency. In a retrospect from the present time, it seems as if one of the causes that led to my conservative position was the extreme exaggeration of some glacialists, who found in glacial erosion a destructive agency competent to accomplish any desired amount of denudation – an opinion from which I recoiled too far. Since the publication of my previous essay I had gradually come to accept a greater and greater amount of glacial erosion in the regions of active ice motion; but in spite of this slow change of opinion, the maximum measure of destructive work that, up to last year, seemed to me attributable to glaciers was moderate; and it was therefore with great surprise that I then came upon certain facts in the Alps and in Norway which demanded wholesale glacial erosion for their explanation. The desire of some years past to revise and extend my former essay then came to be a duty, which it is the object of this paper to fulfil.

My former revision of the problem divided the arguments for glacial erosion under four headings: observations on existing glaciers and inferences from these observations; the amount and arrangement of glacial drift; the topography of glaciated regions; and the so-called argument from necessity – that is, the belief that glaciers must have done this and that because nothing else competent to the task could be found. It is not possible for me at present to review all the new material pertinent to the whole problem; attention can be given here chiefly to a few examples under the third heading.

A GLACIATED VALLEY IN CENTRAL FRANCE

It is evident that, if it were possible to obtain a definite idea of the preglacial topography of a glaciated district, the amount of glacial work might be readily determined as the difference between the preglacial and the present form; independent evidence sufficing to prove that general denudation of the rocky crust in the brief postglacial epoch had been inconsiderable. In the glaciated area of the Central Plateau of France, I had opportunity in January, 1899, of seeing a valley that had been locally modified to a determinate amount by a

Fig. 2.1 The glaciated valley of the Rhue

glacier that once descended northwest from the Cantal along the valley of the Rhue to the junction of the latter with the Dordogne. Outside of the glaciated area, the valleys of the plateau – an uplifted and sub-maturely dissected peneplain, mostly of crystalline rocks – frequently follow incised meandering courses, in which the steep concave slopes are regularly opposed to the gentler convex slopes; the latter being spur-like projections of the uplands, advancing alternatively from one and the other side of the valley. Valleys of this kind are singularly systematic in form, as the result of the combined downward and outward cutting by their streams which, already winding or meandering when the erosion of the valleys began, have increased the width of their meander belt while they deepened their valleys. On entering the glaciated valley of the Rhue, it is found that the regularly descending spurs of the non-glaciated valleys are represented by irregular knobs and mounds, scoured on their up-stream and plucked on the down-stream side; and that the cliffs formed where the spurs are cut off, as in Fig. 2.1, are sometimes fully as strong as those which naturally stand on the opposite side of the valley. The spurs generally remain in sufficient strength to require the river to follow its preglacial serpentine course around them, but they are sometimes so far destroyed as to allow the river to take a shorter course through what was once the neck of a spur. The short course is not for a moment to be confounded with the normal cut-offs through the narrowed necks of spurs, such as are so finely exhibited in the meandering valleys of the Meuse and the Moselle. The short courses are distinctly abnormal features, like the rugged knobs to which the once smooth-sloping spurs are now reduced; they are sometimes narrow gorges incised in the half-consumed spurs, and in such cases, the displacement of the Rhue from its former roundabout course is probably to be explained by constraint or obstruction by ice.

It was thus possible in the valley of the Rhue to make a definite restoration of preglacial form, and to measure the change produced by glaciation. The change was of moderate amount, but it was highly significant of glacial action, for it showed that while a slender, fast-flowing stream of water might contentedly follow a serpentine course at the bottom of a meandering valley, the clumsy, slow-moving stream of ice could not easily adapt itself to so tortuous a path. The more or less complete obliteration of the spurs was the result of the effort of the ice stream to prepare for itself a smooth-sided trough of slight curvature; and if the rocks had been weaker, or if the ice had been heavier, or if the glacial period of the Cantal had lasted longer, this effort might have been so successful as to have destroyed all traces of the spurs. Fortunately the change actually produced, only modified the spurs, but did not entirely destroy them; and their rugged remnants are highly significant of what a glacier can do.

ROCKY KNOBS IN GLACIATED AREAS

On thus generalizing the lesson of the Rhue, it is seen that, just before the complete obliteration of the spurs, some of their remnant knobs may be isolated from the uplands whence these preglacial spurs descended. It is out of the question to regard the ruggedness of such knobs as an indication of small change from their preglacial form, as has been done by some observers. The ruggedness is really an indication of the manner in which a glacier reduces a larger mass to smaller dimensions, by plucking on the down-stream side as well as by scouring on the up-stream side. It is possible that knobs in other glaciated valleys than that of the Rhue may be of this origin; they should then be regarded not as standing almost unchanged and testifying to the incapacity of glacial erosion, but as surviving remnants of much larger masses, standing, like monadnocks above a peneplain, as monuments to the departed greater forms. The two knobs at Sion (Sitten) and the Maladeires, all detached from Mont d'Orge in the upper valley of the Rhône, the hills of Bellinzona in the valley of the Ticino, the rocks of Salzburg where the Salzach emerges from the Alps, and even the Borromeo islands in Lake Maggiore, may perhaps be thus interpreted. Rugged as these knobs may be on the down-stream side, it would be an unreasonable contradiction of the conclusions

Fig. 2.2 Glaciated knobs on the Central Plateau of France

based on observations of many kinds to maintain that their ruggedness was of preglacial origin.

The ice stream from the Cantal at one time expanded sufficiently to flood the uplands bordering the valley of the Rhue, where it produced changes of a most significant kind. The neighboring unglaciated uplands are of systematic form; broad, smoothly arched masses rise, round-shouldered, between the narrow valleys that are incised beneath them; the uplands are as a rule deeply soil-covered, and bare ledges prevail only on the stronger slopes of the young valleys that have been eroded since the peneplain was raised to its present upland estate. But within the glaciated area near the Rhue, the broadly rounded forms of the uplands are replaced by a succession of most irregular rocky knobs, from which the preglacial soils have been well scoured away, as in Fig. 2.2. This seems to be a form most appropriate to glacial action on a surface that had been weathered to variable depths in preglacial time. The ice action sufficed to rasp away the greater part of the weathered material, and to grind down somewhat the underlying rock, often giving the knobs a rounded profile; but it did not nearly suffice to reduce the rocky surface to an even grade. The ice action seems here to have resembled that of a torrent which might sweep away the waste on a flood plain and lay bare and erode the rock ledges beneath; but whose duration was not sufficient to develop a graded floor appropriate to its current.

Another example of this kind seems to occur where the huge glacier of the Inn, escaping from its well-enclosed channel within the mountains, once spread forward in a great fan of ice over the foot-hills at the northern border of the Alps and crept out upon the piedmont plain. The glance that I had at this foot-hill district from a passing train gave me the impression that its ruggedness was much greater than usually obtains along the mountain flanks; as if the rolling hills of preglacial time had been scoured to an increasing roughness by an overwhelming ice-flood that would, if a longer time of action had been permitted to it, have worn down all the inequalities to a smooth, maturely graded floor.

THE VALLEY OF THE TICINO

My first entrance into the Alps last year was from the south by the valley of the Ticino. Thirty-one years before I had followed the same valley and admired its bold sides and its numerous waterfalls; but at that time nothing was noticed that seemed inappropriate to the general idea of the erosion of valleys by their rivers. Thirty years is a long enough time for one to learn something new even about valleys, and on my second visit it was fairly startling to find that the lateral valleys opened on the walls of the main valley of the Ticino 500 ft or more above its floor, and that the side streams cascaded down the steep main-valley walls in which they have worn nothing more than narrow clefts of small depth. This set me wondering, not only as to the meaning of so peculiar an arrangement of valleys and streams, but also as to the reason why so peculiar an arrangement should not have sooner attracted attention as an exceptional characteristic of Alpine topography. Playfair long ago, when describing the relation of side valleys to their trunk, showed clearly that they had 'such a nice adjustment of their declivities that none of them join the principal valley either on too high or too low a level: a circumstance which would be infinitely improbable if each of these vallies were not the work of the stream that flows in it' (1802, 102); yet the whole course of the passing century has hardly sufficed to make full application of this law. So much latitude is usually allowed in the relation of branch and trunk valleys that hundreds of observers,

Fig. 2.3 Val d'Osogna, a hanging lateral valley of the Ticino

many of whom must have been cognizant of Playfair's law, have made no note of the extraordinary exceptions to it that prevail in the glaciated valleys of the Alps. Even the most pronounced advocates of glacial erosion, with a few exceptions to be noted below, have been silent regarding the remarkable failure of adjustment between the declivities of lateral and main glaciated valleys. Indeed, in reviewing the writings of those who have accepted a large measure of glacial erosion, one must be struck with the undue attention that they have given to lake basins and the relative inattention to valleys. This disproportion is probably to be explained as a result of the greater contrast that prevails between a river and a lake than between a river and its branch; it is perhaps for this reason that the attention of geologists and geographers has generally been directed to the origin of lakes rather than to the relation of branch and trunk streams, even when the former cascade from their lateral valleys into the main valley. That glacial erosionists made so little claim for the general deepening of glaciated valleys while they demanded a great deepening of those parts of valleys which have been scoured down to form lake basins, has always seemed to me a difficulty in the way of accepting the demanded measure of lake-basin erosion; and this difficulty was supported by the well-attested observation that the side slopes of glaciated valleys manifest no marked or persistent increase of declivity in passing from above to below the limit of glaciation. If glaciers had scoured out deep lake basins, like those of Maggiore and Geneva, they ought to have significantly deepened the valleys up-stream from the lakes; and if the valleys were thus significantly deepened, it seemed as if their slopes should be steeper below than above the limit of glacial action. The denial of the latter requisite seemed to me to carry with it the denial of the two preceding suppositions.

FEATURES OF STRONGLY GLACIATED VALLEYS

It is true that the uppermost limit of glaciation, *QR*, Fig. 2.4, in Alpine valleys is not attended by a persistent change in the steepness of the valley sides, *AE*, *CJ*; but on descending well within the glaciated

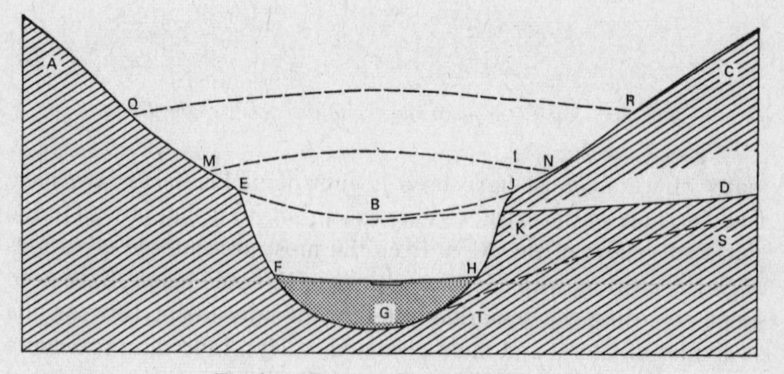

Fig. 2.4 Section of a glaciated valley

valley, a very strong change may usually be found in the slope of the valley walls. The larger valleys, once occupied by heavy glaciers from the lofty central snow fields, are characterized by 'basal cliffs', *EF*, *JH*, that rise several hundred or even a thousand feet above their broad floors, and thus enclose what may be called a 'bottom trough', *EFHJ*, half a mile or a mile wide. The basal cliffs are comparatively straight-walled; they have no sharp spurs advancing into the valley floor. The rock floor, *G*, Fig. 2.4, is buried by gravels, *FH*, to an unknown depth. It is only from the benches above the basal cliffs that the valley sides flare open with maturely inclined slopes; and it is at a moderate depth beneath the level of the benches at the top of these basal cliffs that the lateral valleys, *DK*, open on the walls of the main valley.

The bottom trough within the basal cliffs and beneath the lateral valleys seems to be of glacial origin. It is in the first place a characteristic feature of all the larger glaciated Alpine valleys, as I am assured by Professors Penck, Brückner and Richter, with whom the matter was discussed in the summer of 1899. The non-glaciated valleys manifest no such peculiar form. It is not simply that the terminal portion, *JBK*, of a lateral valley has been cut off by the glacial widening of the main valley floor; the main valley has been strongly deepened, as is assured by the relation of its floor, *FH*, to the prolongation of the floor of the lateral valley, *KB*. The first may be several hundred feet – indeed in some valleys, a good thousand feet – below the second. The lateral valleys must have once entered the main valley at grade, for the flaring sides of the main valley indicate maturity; the side slopes, *AE*, *CJ*, must have once met at *B*. Even the lateral valleys have an open V-section, proving that their streams had cut down to a graded slope, *DB*, that must have led them to an accordant junction with the main river. Nothing seems so competent as glacial erosion to explain the strong discordance of the existing valleys.

The lateral as well as the main valleys have been glaciated, but the former do not exhibit changes of form so distinctly as the latter: in the Ticino system the lateral valleys did not, as far as I saw them, seem to have been much affected by glaciation, a fact that may be attributed to the small size of their branch glaciers in contrast with the great volume of the trunk glacier. There is no sufficient evidence that the valley floor between the basal cliffs has been faulted down, after the fashion of a *graben*; nor is it satisfactory to explain the bottom trough as having been worn out by normal trunk-river erosion, leaving the side streams as it were hanging or suspended above them, for to admit such an origin would be to go counter to all that has been learned regarding the systematic development of valleys. Here it is with regret that I must differ from the opinion of two eminent Swiss geologists who explain the deepening of the main valleys by a revival in the erosive power of the rivers as a result of a regional uplift, while they regard the hanging lateral valleys as not yet accordantly deepened by their smaller streams. It is true that narrow trenches are cut in the floors of the hanging valleys, showing that their streams have made some response to the erosion of the bottom trough in the main valley, and if the bottom trough were a narrow canyon, this relation of trunk and branch streams might be considered normal; but if the breadth as well as the depth of the bottom trough had been acquired by normal river erosion, the side valleys should now, it seems to me, have been trenched much deeper than they are, to some such slope as *ST*, Fig. 2.4.

The opinions of L. Rütimeyer and A. Heim on this question are as follows: Rütimeyer gave an excellent account of hanging lateral valleys thirty years ago in his description of the valley of the Reuss (1869, 13–24). He recognized benches or *Thalstufen* on each side of the valley above the basal cliffs of the existing bottom trough, and regarded them as the remnants of a former, wide open valley floor. Side valleys of moderate fall enter the main valley about at the level of the *Thalstufen*, and their waters then cascade down over the basal cliffs to the Reuss. Glacial erosion is dismissed as incompetent to erode the bottom trough; indeed, the time of glacial occupation of the valley is considered a period of rest in its development. The discordance of main and lateral valleys is ascribed entirely to the differential erosion of their streams. Heim's views on this matter are to be found in his *Mechanismus der Gebirgsbildung* (1878, 1, 282–301) and in an article 'Über die Erosion im Gebiete der Reuss' (1879). He recognizes that the bottom troughs have been excavated in the floors of pre-existing valleys, whose stream lines had been reduced to an even grade (profile of equilibrium, *Gleichgewichtslinie*) and whose lateral slopes had been maturely opened. The side streams must at that time have eroded their valleys deep enough to enter the main valley at accordant grade as stated above. Since then, it is concluded that an elevation of the region has caused a revival (*Neubelebung*) of the main river; and the present greater depth of the main valley is, according to Heim, merely the natural result of this revival, while the smaller side streams have not yet been able to deepen their valleys. The height of the *Thalstufen* or remnants of the former valley floor, seen in the benches above the basal cliffs of the bottom trough, is taken as a measure of the elevation that the mountain mass has suffered.

Apart from the improbability that the deepening of a bottom trough by a revived main river could truncate so many lateral valleys with so great nicety as is repeatedly the case, leaving their streams to cascade down in clefts but slightly incised in the main valley walls, the following considerations lead me to reject the possibility of explaining the discordance between side and main streams by a normal revival of river action.

RELATION OF TRUNK AND BRANCH VALLEYS

The general accordance of maturely developed main and lateral valleys in non-glaciated regions, as recognized by Playfair, is today fully established by innumerable observations in many parts of the world. Truly, during the attainment of mature development, it is possible that a large river may outstrip a small branch stream in the work of deepening its valley, but the discordance thus produced can prevail only during early youth; for as soon as the main river approaches grade the further deepening of its valley is retarded, while at the same time the steepened descent of the lateral streams at their entrance into the main valley accelerates their erosive work. Hence, even if a large trunk river has for a time eroded its valley to a significant depth beneath the tributary valleys, this discordance cannot endure long in the history of the river. It should be noted that discordance of side and main valleys may also be found where a large river has lately been turned to a new path, as in the normal progress of the capture of the upper course of one river by the headward gnawing of a branch of another river (see reference to Russell below), or in the new arrangements of drainage lines in a region from which a glacial sheet has lately withdrawn. Furthermore, the valleys of very small wet-weather streams are frequently discordant with the valley of a serpentine river, if they enter it from the upland that is under-cut by the concave bank of the river. But these cases cannot find application in the hanging valleys of the Alps. The hanging valleys that open on sea cliffs, such as those of Normandy, are of course quite another matter.

OVERDEEPENED MAIN VALLEYS AND HANGING LATERAL VALLEYS

Now it is characteristic of the bottom troughs of the glaciated Alpine valleys that they are broad-floored; they cannot be described as canyons in any proper sense of that word: the walls are steep enough, but they are too far apart. If the existing breadth of the troughs had been acquired in the ordinary manner by the lateral swinging of the main stream and by the lateral weathering of the walls, the long time required for such a change would have amply sufficed for the lateral streams to cut down their valleys to grade with the main valley; and their persistent failure to do so indicates the action of something else than normal river work in the widening of the main valley. This is the very kernel of the problem.

If a main valley were excavated along a belt of weak rocks, the side valley might stand for some time at a considerable height above the main valley floor. Certain hanging valleys in the Alps seem at first sight to belong to this class, but such is not really the case. For example, where the Linth flows into the Wallen See, the well-defined bottom troughs of the river and of the lake both pass obliquely through a syncline of strong lower Cretaceous limestone, which forms cliffs on their walls. Side streams drain the high synclinal areas; one such stream cascades from the west into the Linth trough back of the village of Näfels; another cascades from the north into the Wallen See near its western end. The first explanation for such falls is that they are normally held up on the resistant limestone; but it should be noted that the bottom troughs of the Linth and the Wallen See have been cut down and broadly opened in the same limestones. If the troughs were of normal river origin, the side streams also should have by this time trenched the limestones deeply, instead of falling over the limestone cliffs at the very side of the larger troughs. In the Ticino valley where the side streams are most discordant, massive gneisses prevail; the structure is so nearly uniform over large areas that it affords no explanation of the strong discordance between side and main valleys.

It thus seems obligatory to conclude that the bottom troughs of the larger Alpine valleys were deepened and widened by ice action. This belief is permitted by the abundant signs of glacial erosion on the spurless basal cliffs, and required by the persistent association of over-deepened bottom troughs and discordant hanging lateral valleys with regions of strong glaciation.

SUBAERIAL EROSION DURING THE GLACIAL PERIOD

It should not be imagined that the glacial erosion of troughs in valley floors was necessarily so rapid that no significant subaerial erosion was accomplished during its progress. Ordinary weathering and down-hill transport of rock waste must have been in active operation on the valley sides above the border of the ice-filled channels; and the very fact that on the upper slopes of the mountains, preglacial, glacial and postglacial erosion was similarly conditioned, makes it difficult to distinguish the work done there in each of these three chapters of time. In the diagrams accompanying this article no indication of change from preglacial to postglacial outline on the upper mountain slopes is indicated, because no satisfactory measure can be given to it.

LAKE LUGANO

In the presence of a variety of evidence collected for some years previous to my recent European trip, it had been my feeling that the best explanation offered for the large lakes that occupy certain valleys on the Italian slope of the Alps was that they had resulted from what has been called valley-warping, as set forth by Lyell, Heim and others. It was my desire to look especially at Lakes Maggiore, Lugano and Como with this hypothesis in mind, and to subject it to a careful test by means of certain associated changes that should expectedly occur on the slopes of the neighboring mountains, as may be explained as follows.

On the supposition of moderate or small glacial erosion, a well-matured stage of dissection must have been attained in the district of the Italian lakes in preglacial time; for the main valleys are widely opened, and even the lateral valleys have flaring slopes. In a mature stage of dissection mountains should exhibit a well-advanced grading of their slopes; that is, their sides should be worn back to a comparatively even declivity with little regard to diversity of structure; the descending streams of waste being thus seen to correspond to the flood plains of graded rivers. The agencies of weathering and transport are delicately balanced wherever graded slopes prevail; and a slight tilting of the mountain mass might suffice to disturb the adjustment between the supply and the removal of waste; then all the steepened slopes would soon be more or less completely stripped of their waste cover; their rock ledges would be laid bare, although still preserving the comparatively even declivity that had been gained under the slowly moving waste.

If the lakes had been formed by warping, it is possible to deduce with considerable accuracy the localities where the mountain slopes would be steepened and stripped; namely, the northern slopes about the southern end of the lakes, and the southern slopes about the northern end; but as far as I was able to examine the district about Lake Lugano, no effects of such a warping and tilting were to be detected. The submergence of lateral valleys about the middle of the lakes is also a necessary consequence of the theory of warping; but although the main valley floor is now deep under water, the side valleys are not submerged. Failing to find evidence of warping, and being much impressed with the evidence of deep glacial erosion as indicated by the hanging lateral valleys of the overdeepened Ticino, I examined the irregular troughs of Lake Lugano for similar features, and found them in abundance.

One of the reasons why Lake Lugano had been selected for special study was that it did not lie on the line of any master valley leading from the central Alps to the piedmont plains; hence, if influenced by ice action at all, its basin must have been less eroded than those of Como and Maggiore on the east and west. But in spite of this peculiarity of position, Lugano received strong ice streams from the great glaciers of the Como and Maggiore troughs (see 'Glacial Distributaries', below), and its enclosing slopes possess every sign of having been strongly scoured by ice action. The sides of the lake trough are often steep and cliff-like for hundreds of feet above present water level, thus simulating the basal cliffs of the Ticino valley; while at greater heights the valley sides lean back in relatively well-graded slopes. The angle at the change of slope is often well defined, but it is independent of rock structure. Narrow ravines are frequently incised in the basal cliffs, and alluvial fans of greater or less size are built into the lake waters from the base of the ravines.

The northeastern arm of the lake, extending from the town of Lugano to Porlezza, receives several cascading streams from hanging valleys on its southern side. The side slopes of the hanging valleys are for the most part flaring open and well graded, from which it must be concluded that their streams had, under some condition no longer existing, ceased to deepen their valleys for a time long enough to allow the valley sides to assume a mature expression; and that since then the bottom trough of the main arm of the lake has been eroded deep and wide, with a very small accompanying change in the lateral valleys. In other words, the side valleys were, in preglacial time, eroded to a depth accordant with the floor of the master valley that they joined, and since then the bottom trough has been eroded in the floor of the master valley by a branch of the Como glacier. In post-glacial time the side streams have begun to trench their valley floors, eroding little canyons; but much of this sort of work must be done before the side valleys are graded down even to the level of the lake waters, much less to the level of the bottom of the lake.

The two southern arms of the lake lead to troughs whose floors ascend southward to the moraines of the foot-hills, beyond which stretch forward the abundant overwashed gravels of the great plain of the Po.

I do not mean to imply that every detail of form about Lake Lugano can find ready explanation by the mature glacial modification of a mature preglacial valley system; but a great number of forms may be thus explained, and a belief in strong glacial erosion was forced upon me here as well as in the valley of the Ticino. A detailed study of the Italian lakes with the intention of carefully sorting out all the glacial modifications of preglacial forms would be most profitable.

VARIOUS EXAMPLES OF GLACIATED VALLEYS

My excursions of last summer showed me a number of over-deepened main valleys and hanging lateral valleys in the Alps; for example, those of the Inn and of the Aar. Lakes Thun and Brienz receive numerous cascades from hanging valleys that stand high above the water surface. The valley of Lauterbrunnen also affords a conspicuous illustration of a deep bottom trough enclosed by high basal cliffs that rise to the edge of more open upper slopes; the celebrated Staubbach fall is the descent of a small lateral stream from its lofty hanging valley, and the picturesque village of Mürren, *M*, Fig. 2.5, stands on the flaring slope or *Thalstufe* of the preglacial valley, just above the great basal cliff of glacial origin. A mile or so south of the village of Lauterbrunnen, the Trummelbach, *T*, Fig. 2.5, descends the precipitous eastern wall from a hanging valley whose floor is hundreds of feet above that of the Lütschine; it is roughly sketched in Fig. 2.6. Although the lateral Trummelbach brings a large volume of water to the main valley, it descends by a very narrow cleft in the rock face, a trifling incision in the valley wall; while the main valley, whose trunk stream did not seem to be more than five times the volume of its branch, is half a mile or more broad, wide open and flat-floored. The

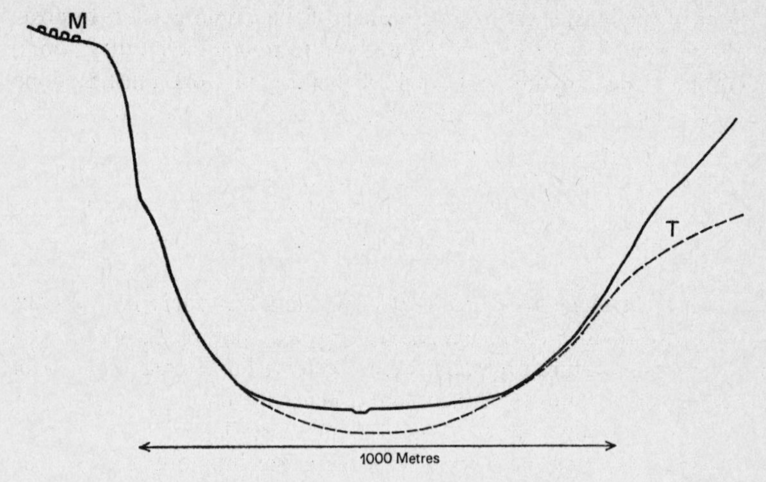

Fig. 2.5 True-scale cross-section of the Lauterbrunnen valley

cross-section of the main valley is over a thousand times as large as that of the lateral cleft. Such a disproportion of main valley and lateral cleft is entirely beyond explanation by the inequality of their streams; and for those who feel that they must reject glacial erosion as the cause of the disproportion, there seems to be no refuge but in ascribing the main valley to recent down-faulting: a process that can hardly be called on to follow systematically along the floors of the larger glaciated valleys of the Alps, and to avoid the non-glaciated valleys and the mountain ridges.

Certain well-known Alpine glaciers may be instanced as reaching just beyond the end of a hanging lateral valley and thence cascading into the deeper main valley. One is the Mer de Glace by Chamounix; its cascading end is known as the Glacier des Bois. Another is the neighboring Glacier des Bossons, from whose upper amphitheatre a steep tongue descends far below; like the waterfalls of Norway, the tongue may be seen lying on the side slope from a considerable distance up or down the main valley. A third example is the Glacier of the Rhône, whose splendid terminal cascade is so conspicuous from the road to the Furka pass. Possibly the Vernagt glacier is another of the same kind; its catastrophic overflows into the lower Rofen valley have often been described. Doubtless many other examples of this class might be named.

Fig. 2.6 Diagram of the gorge of the Trummelbach, Lauterbrunnen valley

While engaged upon these observations in the Alps in the spring of 1899, I sent a brief note about them to my esteemed friend, Mr G. K. Gilbert of Washington, telling him that all the lateral valleys seemed to be 'hung up' above the floors of the trunk valleys. His reply was long in coming to Europe, and, on arriving at last, it was dated Sitka, Alaska, where Mr Gilbert had gone as a member of the Harriman Alaskan Expedition, and where my note had been forwarded. He wrote that, for the fortnight previous to hearing from me, he and his companions had been much impressed with the discordant relations of lateral valleys over the waters of the Alaskan fiords, and he suggested that such laterals should be called 'hanging valleys' – a term which I have since then adopted. He fully agreed that hanging valleys presented unanswerable testimony for strong glacial erosion, as will be stated in his forthcoming report on the geology of the Expedition.

After leaving Switzerland, I had a brief view of the lake district in northwest England, before crossing to Norway. The amount of glacial erosion in the radiating valleys of the English lakes has been much discussed, and, as usual, directly opposite views have been expressed. Rugged rocky knobs were seen in abundance about Ambleside and along the ridge separating the valley of Thirlmere from St John's Vale; and the latter receives a hanging valley from the east near Dalehead post-office. The famous falls of Lodore seemed to descend from the mouth of a hanging valley into Derwentwater.

FIORDS AND HANGING VALLEYS IN NORWAY

In Norway I had the pleasure of making a ten days' cross-country excursion in company with Dr Reusch, Director of the Norwegian Geological Survey. We entered from Bergen through Hardanger fiord, and crossed the highlands by the Haukelisaetr road to Skien on the southeastern lowlands, thus making a general cross-section on which many characteristic features were seen. Norway has long been known as a land of waterfalls, but it is not generally stated with sufficient clearness or emphasis that many or most of the falls are formed by the descent of streams from maturely opened trough-like hanging valleys which are abruptly cut off by the walls of the fiords. The discordance between main and side streams is simply amazing. The fiord valleys are frequently one or two miles wide; the waters of the fiords are of great depths, reaching 3000 ft in some cases. Even when a side valley stands but little above sea-level, its floor may be half a mile above the floor of the fiord. On passing inland beyond the head of the fiord water, where the whole depth of the fiord valley is visible, the side valleys may open more than a thousand feet above the main valley floor. In many cases where the fiords are enclosed by smooth walls, the cascading side streams have not yet incised a cleft in the bare rock surface, so that their foaming waters are visible for many miles up and down the fiord. Streams of considerable size sometimes plunge down from the rolling uplands in whose edge they seem to have just begun to cut a cleft. Abnormal discordance of trunk and branch stream is, therefore, a strongly marked characteristic of the Norwegian drainage. The necessity for appealing to strong glacial erosion in explanation of this prevailing discordance, may be set forth as follows.

MEASURE OF GLACIAL EROSION IN NORWEGIAN FIORDS

The deep valleys of Norway, partly occupied by sea water, are incised beneath an uneven highland which bears so many hills and mountains that it makes little approach to a peneplain, yet which here and there shows so many broadly opened uplands between the hills and mountains that it may be taken to represent the well-advanced work of a former cycle of denudation when the region stood much lower than it stands now. As a whole, a mature or late mature stage seems to have been reached before a movement of uplift introduced the present cycle. Let us now make two suppositions regarding the work of normal river erosion in the preglacial part of the present cycle, in order to determine, if possible, how much additional erosion must be attributed to ice in the production of existing forms.

First, let it be supposed that the revived main rivers had incised their valleys to the depth of the present fiords in preglacial time, and that the discordance of main and side valleys now visible is the appropriate result of the youth of the present cycle. If we recall only the steepness of the fiord walls, this supposition might be justified, and thus the amount of glacial erosion needed to develop existing forms would be small. But it must not be forgotten that the fiords, although often steep-walled, are always broad, much broader than a young preglacial valley could have been at that stage of early youth when its side streams had not cut down to its own depth. Hence glacial erosion must, under this supposition, be appealed to for the widening of preglacial canyons, steep-walled and narrow, into the existing fiord troughs, steep-walled and broad. At the middle of the fiord troughs, the lateral erosion thus demanded would often measure thousands of feet, and that in the most massive and resistant crystalline rocks.

A second supposition leads to no greater economy of glacial action. Let it be supposed that the revived streams of preglacial time had reached maturity before the advent of the glacial period. In that case, the side streams must have entered the main streams at accordant grade, and hence the main valleys could not then have been cut much deeper than the side valleys are now cut; not so deep, indeed, for the side valleys have been somewhat deepened by glacial action, if one may judge by their trough-like form as well as by the evidence of intense glacial action all over the uplands, even over most of the surmounting hills and mountains. Hence, to develop the existing discordant valley system from a mature preglacial valley system of normal river erosion, requires a great deepening of the fiords by ice action, again to be measured in thousands of feet. Thus there seems to be no escape from the conclusion that glacial erosion has profoundly modified Norwegian topography. As far as I could judge from my brief excursion over the highlands, either one of the two suppositions above considered is permissible, provided only that strong glacial erosion comes after the river work of the current cycle.

If the Hardanger fiord may be taken as the type of its many fellows, one may say that hanging lateral valleys are the rule, not the exception, in Norway. Furthermore, the smoothed, spurless walls of the larger fiords, composed of firm bare rock from the upland to water edge, do not resemble the ravined and buttressed sides of normal valleys. The marks of downward water erosion are replaced by what seem to be marks of nearly horizontal plucking and scouring. Blunt-headed valleys and corries (*botner*) both seem beyond production by normal weathering and washing. Yet, striking as these features are, they do not seem to me so compulsory of a belief in strong glacial erosion as the hanging valleys that have so little relation to the fiords beneath them, and the flaunting waterfalls that descend so visibly from the hanging valleys, instead of retiring, as is the habit of falls all over the unglaciated parts of the world, into ravines where they are hid to sight from most points of view.

The rocky islands that rise from the shallower parts of the fiords should not be taken as signs of feeble glacial erosion, but rather as remnants surviving from the destruction of larger masses in virtue of some slight excess of resistance. A well-known example of this kind is near Odde at the head of the large southern arm (Sörfjord) of the upper Hardanger fiord; in the same neighborhood are several fine hanging valleys, one of which with its cascading stream, the Strandfos, descends into Sandven Lake, just south of the side valley occupied by the well-known Buer glacier.

THE CYCLE OF GLACIAL DENUDATION

The points of resemblance between rivers and glaciers, streams of water and streams of ice, are so numerous that they may be reasonably extended all through a cycle of denudation. Let us then inquire if glaciers may not, during their ideal life history, develop as orderly a succession of features as that which so well characterizes the normal development of rivers. Let us here consider the life history of a glacier under a constant glacial climate, from the beginning to the end of a cycle of denudation, just as I. C. Russell has considered the 'life history of a river' under a constant pluvial climate, in his *Rivers of North America.* Thus young glaciers will be those which have been just established in courses that are consequent upon the slopes of a newly uplifted land surface; mature glaciers will be those which have eroded their valleys to grade and thus dissected the uplifted surface; and old glaciers will be those which cloak the whole lowland to which the upland has been reduced, or which are slowly fading in the milder climate of the low levels appropriate to the close of the cycle of denudation.

Imagine an initial land surface raised to a height of several thousand feet, with a moderate variety of relief due to deformation. Let the snow line stand at a height of 200 ft. As elevation progresses, snow accumulates on all the upland and highland surfaces. Glaciers are developed in every basin and trough; they creep slowly forward to lower ground, where they enter a milder climate (or the sea) and gradually melt away. At some point between its upper heads and its lower end, each glacier will have a maximum volume. Down stream from this point, the glacier will diminish in size, partly by evaporation but more by melting; and the ice water thus provided will flow away from the end of the glacier in the form of an ordinary stream, carving its valley in normal fashion. Some erosion may be accomplished under the upper fields of snow and névé, but it is believed that more destructive work is done beneath the ice. The erosion is accomplished by weathering, scouring, plucking and corrading. Weathering occurs where variations of external temperature penetrate to the bed-rock, as is particularly the case between the séracs of glacial cascades, and again along the line of deep crevasses or bergschrunds that are usually formed around the base of reservoir walls, which are thus transformed into corries (cirques, karen, botner) as has been suggested by several observers; scouring is the work of rock waste dragged along beneath the glacier, by which the bed-rock is ground down, striated and smoothed; plucking results from friction under long-lasting heavy pressure, by which blocks of rock are removed bodily from the glacier bed and banks; corrading is the work of subglacial streams, which must be well charged with tools, large and small, and which must often flow under heavy pressure and with great energy. All these processes are here taken together as 'glacial erosion'.

Let it be assumed that at first the slope of a glacier's path was steep enough to cause it to erode for the greater part or for the whole of its length. Each young glacier will then proceed to cut down its valley at a rate dependent on various factors, such as depth and velocity of ice stream, character of rock bed, quantity of ice-dragged waste, and so on; and the eroded channel in the bottom of the valley will in time be given a depth and width that will better suit the needs of ice discharge than did the initial basin or trough of the uplifted surface. The upper slopes of the glacial stream will thus be steepened, while its lower course will be given a gentler descent. Owing to the diminution of the glacier toward its lower end, the channel occupied by it will diminish in depth and breadth downwards from the point of maximum volume; this being analogous to the decrease in the size of the channel of a withering river below the point of its maximum volume. A time will come when all the energy of the glacier on its gentler slope will be fully taxed in moving forward the waste that has been brought down from the steeper slopes; then the glacier becomes only a transporting agent, not an eroding agent, in its lower course. This condition will be first reached near the lower end, and slowly propagated headwards. Every part of the glacier in which the balance between ability to do work and work to be done is thus struck may be said to be 'graded'; and in all such parts, the surface of the glacier will have a smoothly descending slope. Maturity will be reached when, as in the analogous case of a river, the nice adjustment between ability and work is extended to all parts of a glacial system. In the process of developing this adjustment, a large trunk glacier might entrench the main valley more rapidly than one of the smaller branches could entrench its side valley; then for a time the branch would join the trunk in an ice-rapid of many séracs. But when the trunk glacier had deepened its valley so far that further deepening became slow, the branch glacier would have opportunity to erode its side valley to an appropriate depth, and thus to develop an accordant junction of trunk and branch ice *surfaces*, although the *channels* of the larger and the smaller streams might still be of very unequal depth, and the channel *beds* might stand at discordant levels. If the glaciers should disappear at this stage of the cycle, their channels would be called valleys, and the discordance of the channel beds might naturally excite surprise. The few observers who, previous to 1898, commented upon a discordance of this kind, explained it as a result of excessive erosion of the main valley by the trunk glacier; while the hanging lateral valleys were implicitly, if not explicitly, regarded as hardly changed from their preglacial form.

When the trunk and branch glaciers have developed well-defined, maturely graded valleys, the continuous snow mantle that covered the initial uplands of early youth is exchanged for a discontinuous cover, rent on the steep valley sides where weathering comes to have a greatly increased value, and thickened where the ice streams have established their courses. This change corresponds to that between the ill-defined initial drainage in the early youth, and the well-defined drainage in the maturity, of the river cycle.

It is probable that variations in rock structure will have permitted a more rapid development of the graded condition in one part of the glacial valley than in another, as is the case with rivers of water. Steady-flowing reaches and broken rapids will thus be produced in the ice stream during its youth; and the glacial channel may then be described as 'broken-bedded'. But all the rapids must be worn down and all the reaches must become confluent in maturity. It is eminently possible that the reaches on the weaker or more jointed rocks may be eroded during youth to a somewhat greater depth than the sill of more resistant or less jointed rock next down stream; and if the glacier should vanish by climatic change while in this condition, a lake would occupy the deepened reach, while the lake outlet would flow forward over rocky ledges to the next lower reach or lake. Many Norwegian valleys today seem to be in this condition. Indeed some observers have described broken-bedded valleys as the normal product of glacial erosion, without reference to the early stage in the glacial cycle of which broken-bedded glacial channels seem to be characteristic. Truly, it is not always explicitly stated that the resistance of the rock bed varies appropriately to the change of form in a broken-bedded channel; but the variations of structural resistance or firmness that the searching pressure and friction of a heavy glacier could detect might be hardly recognizable to our superficial observations; and on the other hand the analogy of young ungraded glaciers with young ungraded rivers seems so natural and reasonable that broken-bedded glacial channels ought to be regarded only as features of young glacial action, not as persistent features always to be associated with glacial erosion. If the glaciers had endured longer in channels of this kind, the 'rapids' and other inequalities by which the bed may be interrupted must have been worn back and lowered, and in time destroyed.

If a young glacier erodes its valley across rocks of distinctly different resistances, a strong inequality of channel bed may be developed. Basins of a considerable depth may be excavated in the weaker strata, while the harder rocks are less eroded and cross the valleys in rugged sills. Forms of this kind are known in Alpine valleys; for example, in the valley of the Aar above Meiringen and in the lower Gasternthal near its junction with the Kanderthal; in both these cases the basins have been aggraded and the sills have been trenched by the postglacial streams. In the lower Gasternthal the height and steepness of the rocky sill, when approached from up-stream, is astonishing; its contrast to the basin that it encloses is difficult enough to explain even for those who are willing to accept strong glacial erosion. It should, however, be noted that river channels also are deeper in the weaker rocks up-stream from a hard rock sill; if the river volume should greatly decrease, a small lake would remain above the sill, drained by a slender stream cutting a gorge through the sill.

If an initial depression occurred on the path of the glacier, so deep that the motion of the ice through it was much retarded, an ice-lake would gather in it. Then the waste dragged into the basin from up-stream might accumulate upon its floor until the depth of the basin was sufficiently decreased and the velocity of the ice through it sufficiently increased to bring about a balance between ability to do work and work to be done. Here the maturely graded condition of the ice stream would have been attained by aggrading its bed, instead of degrading it; this being again closely analogous to the case of a river, which aggrades initial depressions and degrades initial elevations in producing its maturely graded course.

Water streams subdivide toward the headwaters into a great number of very fine rills, each of which may retrogressively cut its own ravine in a steep surface, not cloaked by waste. But the branches of a glacial drainage system are much more clumsy, and the channels that they cut back into the upland or mountain mass are round-headed or amphitheatre-like; but the beds of the branching glaciers cannot be cut as deep as the bed of the large glacial channel into which they flow: thus corries, perched on the side-walls of large valleys, may be produced in increasing number and strength as glacial maturity approaches, and in decreasing strength and number as maturity passes into old age. As maturity approaches, the glacial system will include not only those branches that are consequent upon the initial form, but certain others which have come into existence by the head-ward erosion of their névé reservoirs following the guidance of weak structures; thus a maturely developed glacial drainage system may have its subsequent as well as its consequent branches. It is entirely conceivable that one ice stream may capture the upper part of another. The conditions most favorable for such a process resemble those under which river diversions and adjustments take place; namely, a considerable initial altitude of the region, allowing a deep dissection; a significant difference of drainage areas or of slopes, whereby certain glaciers incise deeper valleys than others; a consider-able diversity of mountain structure, permitting such growth and arrangement of subsequent glaciers as shall bring the head reservoir of a subsequent ice stream alongside of and somewhat beneath the banks of a consequent ice stream. Thus glacial systems may come to adjust their streams to the structures upon which they work, just as happens in river systems.

As the general denudation of the region progresses, the snow fall must be decreased and the glacial system must shrink somewhat, leaving a greater area of lowland surface to ordinary river drainage. When the upland surface is so far destroyed that even the hill tops stand below the 200-ft contour, the snow fields will be represented only by the winter snow sheet, and the glaciers will have disappeared, leaving normal agencies to complete the work of denudation that they have so well begun.

If a snow line at sea-level be assumed, glaciation would persist even after the land had been worn to a submarine plain of denudation at an undetermined depth beneath sea-level. The South Polar regions offer a suitable field for the occurrence of such a surface.

Whether glaciers of the Norwegian or of the Alpine type shall occur, is dependent partly on initial conditions, partly on the stage of advance through the cycle of denudation. If the initial form offer broad uplands, separated by deep valleys, snow fields of the Nor-wegian type may have possession of the uplands during the youth of the glacial cycle; but when maturity is reached, the uplands will be dissected, and the original confluent snow field will be resolved into a number of head reservoirs, separated by ridges. On the other hand, as the later stages of the cycle are approached, the barriers between adjacent reservoirs will be worn away, and they will tend to become confluent, here and there broken only by nunataks. If the snow line lay low enough, a completely confluent ice and snow shield would cover the lowland of glacial denudation when old age had been reached. If the glacial conditions of Greenland preceded as long as they have followed the glacial period over the rest of the North Atlantic region, who can say how far the ice of the Greenland shield has modified the forms on which its work began!

GLACIAL DISTRIBUTARIES

If a maturely dissected mountain range were occupied by snow-fields and glaciers of large size, certain peculiar results might be expected near the mountain base. Under normal preglacial conditions, a small low ridge suffices for the complete separation of two river systems, because the channels of rivers are so small in comparison to their valleys. But glacial channels are a large part of their valleys, and when great glaciers from the lofty mountain centres descend by the master valleys to the mountain flanks or even to the piedmont plains, distributary ice streams or outflowing branches may naturally enough be given off wherever the ice surface rises high enough to overtop the ridges by which the master valleys are separated from adjacent minor valleys. If a distributary branch has sufficient strength and endurance, it may wear down the ridge that it crosses and thus increase and perpetuate its lateral discharge; but it cannot usually be expected to erode a channel as deep as that of the main glacier from which it departs. On the disappearance of the ice, a hanging valley will be left above the floor of the master valley; but in this case, the drainage of the hanging valley will be away from, not toward, the master. Here we probably have the explanation of those broad hanging valleys which lead from the valley of Lake Maggiore on the west and, less distinctly, from that of Lake Como on the east to the compound basin of the intermediate Lake Lugano. On going southward by rail from Bellinzona to Lugano, along a stretch of the St Gotthard route between the great tunnel and Milan, the railway obliquely ascends the southeastern wall of the trough-like valley of the Ticino just above the head of Lake Maggiore; and at a height of several hundred feet over the delta flood-plain the line turns off to a well-marked hanging valley in which the stream runs away from the Ticino to Lake

Lugano. The notch made by this supposed glacial distributary is a conspicuous feature in the view from Bellinzona and thereabouts.

The anomalous forking of Lake Como and the open branch from the main valley of the Rhine at Sargans through the trough of Wallen See to Lake Zurich appear to be the paths of large glacial distributaries which eroded their channels deeply across divides that presumably existed in preglacial time. The west wall of the main valley of the Isère in the Alps of Dauphiny, southeastern France, is deeply breached by passes that lead northwest to the troughs of Lakes Annecy and Bourget, through which the distributaries of the Isère glacial system must have flowed.

It may be further supposed that if the preglacial valleys were so arranged that a glacial distributary found a shorter and steeper course to the piedmont plain or to the sea than that followed by the master glacier, the distributary might under a long enduring glaciation become the main line of glacial discharge; and if so, it could be eroded to a greater depth than the former master valley at the point of divergence. In such a case, the postglacial river drainage would differ significantly from the preglacial.

THE DEPTH OF MATURE GLACIAL CHANNELS

The depth with respect to sea-level to which the channels of a glacial system may be eroded when the graded condition is reached, is a subject of special interest. For many miles along the lower course of a branchless trunk glacier, its volume is lessened by melting and evaporation, and at its end the ice volume is reduced to zero; slow ice motion being progressively replaced by rapid water motion. In such a case the law of continuity does not demand that the ice velocity shall be inversely proportional to the area of the cross section, as is the case in the normal river (where it is assumed that there is no loss by evaporation). Indeed, in the lower trunk of a mature glacier, it may well be that the velocity of ice movement is in a rough way directly proportional to cross-section area. This appears to be verified by measurements of the Rhône glacier, where the mean annual movement is 110 m in the heavy trunk above the cascade, 27 m just below the cascade, and only 5 m close to the melting front. Evidently then, the erosion of the glacial bed, in so far as it is determined by the pressure and motion of the ice stream, will have its maximum some distance up-stream from the end of the glacier (J. Geikie, 1898, 236).

The glacial channel must therefore become narrower and shallower as its end is neared, as has already been stated. If the glacier ends some distance inland from the sea, its action will be conditioned by the grade and length of the river that carries away the water and waste from its lower end. The deepening of the distal part of the channel accomplished in youth might be followed by a shallowing for a time during maturity, when the accumulation of morainal and washed materials in front of the glacier compelled its end to rise. Now it may well be conceived that the surface slope of such a glacier near its end is less than the angle between the surface and the bottom of the glacier; and in this case, the glacial floor must become lower and lower for a certain distance up-stream. If such a glacier should melt away, the distal part of its channel would be occupied by a lake, although even the head of the lake may not reach to the locus of maximum glacial erosion. Lakes Maggiore, Como and Garda seem to occupy basins whose distal enclosure by heavy moraines and sheets of over-washed gravels has added to the depth produced by erosion further up-stream. It would seem, however, that a lake basin thus situated must be only a subordinate incident in the general erosion of the whole length of the glacial channel. Too much attention has, as a rule, been given to lakes of this kind, and not enough to the other effects of glacial action; it seems especially out of proportion to suppose that the maximum erosion by a glacier takes place near its end, as has been done by some authors, on account of the prevalent occurrence of lakes in this situation.

If a glacier advances into the sea and ends in an ice cliff, from which ice blocks break off and float away, something of a basin-like form of its lower channel may be produced; but the dimensions of this basin will be determined by the climate at the termination of the glacier. If the climate is such as to allow the glacier to enter the sea in maximum volume, then a basin is not to be expected. The more the glacier diminishes towards its end, the less erosion and the more deposition may occur beneath it, and the more of a basin may be developed inland from its end.

The depth to which a glacier may cut its channel when it enters the sea is of particular importance. If the glacier is 1000 ft thick at its end, it must continue to press upon and scour its bed until only about 140 ft of ice remain above sea-level; its channel will thus be worn more than 800 ft beneath sea-level. Truly, the latter part of this work will be performed with increasing slowness; but if time enough be allowed the work must be accomplished, just as is the case with rivers. If a glacier should melt away from its deep entrenchment, its channel would be occupied by an arm of the sea or fiord, reaching many miles into the land. The fiord might be shallower at its mouth than further inland, if differential erosion and deposition had occurred along its channel.

An important corollary from this conclusion – perhaps not so much of a novelty to glacial erosionists as to their confrères of the opposite opinion – is that the depth of water in the fiords of a strongly glaciated coast is not a safe guide to the movement of the land since preglacial time. If there had been a still-stand of the earth's crust through the whole glacial period, the preglacial river channels that were graded down a little below sea-level at their mouths would be replaced by glacial channels that might be eroded hundreds of feet below sea-level. The depth of fiords thus seems to depend on the size of their ancient glaciers, on the height of the mountain background, and on the duration of the glacial period, as well as on movements of the land. If liberal measures of glacial erosion and glacial time are allowed, no depression of glaciated coasts since preglacial time is needed to account for their peculiar features. The glacial channels may have been simply invaded by the sea, as the ice melted away, without any true submergence.

THE ORIGIN OF CORRIE BASINS

On pursuing the above line of consideration a little further, it may give some light on the occurrence of the small rock basins that are so often found in the floor of cliff-walled corries. Imagine that a large glacial system has become maturely established, and that it 'rises' in many blunt head-branches that have excavated corries in a preglacial mountain mass, and have cut down channels, at their junction with the larger branches or trunk glacier, to a depth appropriate to their volume. Unless the erosion of the corries has been guided by differences of rock structure, there does not seem to be reason for their possessing a basined floor at this stage of development; but if a change of climate should now cause the trunk glacier to disappear, while many of the blunt head-branches remain in their corries, each little glacier thus isolated will repeat the conditions of erosion above inferred for the trunk glacier; and if this style of glaciation linger long enough, rock basins may very generally characterize the floors of the corries when the ice finally melts away. Fig. 2.7 may make this clearer. Let the broken line, *ABC*, be the slope of a preglacial lateral ravine which reaches a trunk stream at *C*, while *ADC* is the profile of an adjoining lateral spur. After vigorous and mature glaciation, the dotted line, *GE*, may represent the surface slope of a lateral glacier, and *GHJ* that of the lateral glacier bed; while *EFL* is the surface of the trunk glacier, and *EKL* the bed. The lower part of the lateral spur has been cut off to make the basal cliff beneath *D*. On the disappearance of the trunk glacier at this stage, the shrunken side

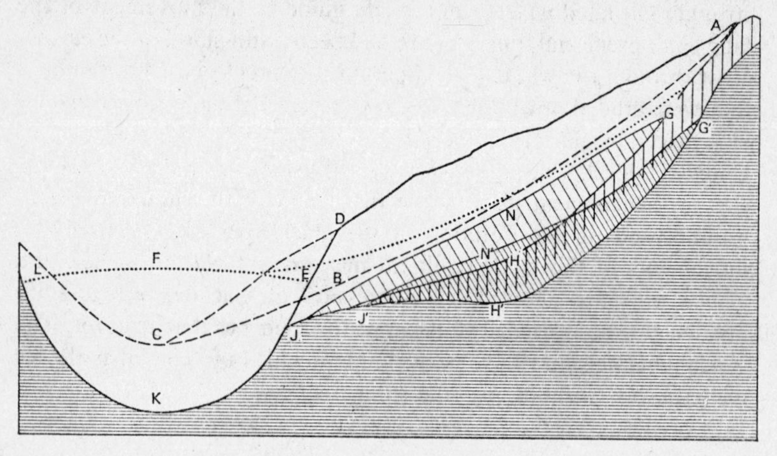

Fig. 2.7 Diagram-section of a lateral valley with a corrie basin

glacier, *GNJH*, occupies its corrie or hanging valley, which opens at *J* on the oversteepened wall, *DJK*, of the evacuated channel of the trunk glacier. Let the maximum erosion of the corrie glacier, as conditioned by pressure and motion, be at *H*. Then after some time the weathering of the cliff walls and the erosion of the floor will have transformed the corrie and its glacier to a form, *G'N'J'H'*, such that the deepening of the glacial bed should be a maximum at *HH'*. The continuous slope of the glacial bed, *GHJ*, appropriate to the time when the lateral glacier joined the trunk glacier, may thus be transformed into a basined curve, *G'H'J'*, appropriate to a small glacier terminating at *J'*; and on the disappearance of the small glacier, a tarn or rock-basin lake may occupy the depression at *H'*. Richter's supposition (1896) that the uplands of Norway result from the consumption of pre-existent mountains by the great extension of corrie-glacier floors, each similar to *J'H'*, seems mechanically possible; but it is nevertheless climatically very improbable, and it seems to me deficient in not attributing enough work to normal preglacial erosion.

PRACTICAL UTILITY OF THE IDEAL GLACIAL CYCLE

In every case, the full understanding of the conditions developed by any system of glaciers, existing or extinct, can be reached only by a complete analysis of the conditions under which they began to work, of the energy with which they worked, of the part of a cycle during which they worked, and of the complications of climatic change or of crustal movements by which their work was modified in this way or in that. A partial analysis may suffice for a particular instance; but the explorer will be better equipped for the explanation of all the instances that he discovers if he sets out with a well-elaborated conception of the ideal glacial cycle of denudation, and of the complications it is likely to suffer. However extensive and definite this conception may be, exploration will probably require its further extension and definition; however brief exploration may be, it will probably be aided by an orderly examination of all pertinent knowledge previously accumulated.

As a practical instance of the value of the glacial cycle, we may consider the aid given toward the solution of certain problems by the careful reconstruction – or at least the conscious attempt at reconstruction – of the form of the land surface on which the Pleistocene glaciers began their work, and by the legitimate deduction of the characteristics of maturity in the cycle of glacial erosion. Beyond the mature stage, we may seldom have occasion to go, as there do not seem to be actual examples of more advanced glacial work. The initial form on which Pleistocene glacial action began is in no case known to be that implied in the opening paragraphs of the section on the Glacial Cycle; namely, a land mass freshly uplifted from beneath the sea and not previously carved by the streams of an ordinary or normal cycle of erosion. In central France, for example, the initial form was an uplifted and submaturely dissected peneplain, in which valleys with incised meanders had been habitually developed. It was there of the greatest assistance to carry into the glaciated area a clear picture of its preglacial form, as determined by generalizing the adjacent non-glaciated area. At the same time, the ideal picture of a maturely developed glacial drainage system, with smooth-sided troughs, was seen to represent a much more advanced condition than was attained in the rugged valley of the Rhue; and thus a tolerably definite idea was gained of the youthful stage of glacial development, somewhere between its beginning and its maturity, and of the amount of destructive work needed to reach this youthful stage. This elementary example illustrates a method embodying the cycle of glacial denudation that ought to be applied whenever possible.

The larger Norwegian fiords may be instanced as glacial channels that present every appearance of having advanced far toward the mature stage of a cycle of glacial denudation from an initial or preglacial form not yet well understood. The variation of form between the main fiords and their branches gives some indication that the glacial work was accomplished in several successive epochs, with the interglacial epochs of normal river work between; but this is only a suggestion, needing much more field work before it can be assured. Not only the deep fiords, but the hanging valleys and the uplands also, have been ice-scoured; for hanging valleys frequently have a well-defined U-section, and sometimes receive secondary hanging valleys from the enclosing uplands; and the streams of the uplands exhibit repeated departures from the forms of normal erosion. Although possessing little drift, the uplands frequently bear lakes of moderate depth and irregular outline; in spite of the breadth to which the upland valleys are opened between the surmounting hills and mountains, their streams frequently change from wandering at leisure in split or braided channels along broad floors, to dashing down in haste over rocky rapids: a behavior that is manifestly inconsistent with that of the mature drainage of a normally denuded region. Even the surmounting hills exhibit strong scouring on their up-ice-stream side. It does not therefore seem permissible to conclude that the hanging valleys which open on the walls of the greater fiords have not been deepened by ice erosion because they escaped the more severe glaciation that scoured out the fiords themselves. All the valleys have been glaciated, and all have been significantly modified from their preglacial form. The discordance of overdeepened main fiord and hanging lateral valley seems to me best explained as the result of the mature development of glacial drainage, in which the chief trunks and the larger branches of the glacial systems had for the most part reached a graded condition. Trunk and branch glaciers would then have united at even grade as to their upper surface, and the trunk and branch channels would have had dimensions satisfactory to the ice currents which flowed through them, but the channel beds would have been discordant, as they are found to be.

LITERATURE

(*See also Bibliography at end of book*)

BARRETT, R. L. (1900) 'The Sundal drainage system in central Norway', *Bull. Am. geogr. Soc.*, XXXII.

BLANFORD, W. T. (1900) 'On a particular form of surface, apparently the result of glacial erosion, seem on Loch Lochy and elsewhere', *Quart. J. geol. Soc.*, LVI 198–204.

BOULE, M. (1896) 'La topographie glaciaire en Auvergne', *Ann. Géogr.*, V 277–96.

BRÜCKNER, E. (1885) 'Die Vergletscherung des Salzachgebietes', *Geogr. Abh., Vienna*, I 1–183.

DAVIS, W. M. (1900) 'Glacial erosion in the valley of the Ticino', *Appalachia*, IX 136–56.

FORBES, J. D. (1853) *Norway and its Glaciers*. Edinburgh.

FOREL, F. A. (1897) 'Fleuves et glaciers', *Bull. Soc. vaud. Sci. nat.*, XXXIII 202–4.

GANNETT, H. (1898) 'Lake Chelan', *Nat. geogr. Mag.*, IX 417–28.

GEIKIE, J. (1895) *The Great Ice Age*. New York. 3rd ed.

—— (1898) *Earth Sculpture*. London.

3 The Profile of Maturity in Alpine Glacial Erosion

WILLARD D. JOHNSON

THE literature of glaciation has not escaped the blemish of too free generalization. It was early asserted of the Sierra Nevada, for example, that Pleistocene glaciers of the alpine type, descending from an ice-cap in the summit region of the range, had reached to the range foot, and that the abnormally large canyons, particularly of the western flank, were the products, from head to foot, of glacial erosion. Such a statement made of southeastern Alaska, of the Scandinavian peninsula, or of the Patagonian Andes would not on the face of it be absurd. Nor was it absurd of the Sierra Nevada. It was possible, despite the low latitude, that ice-streams should have descended to the range foot, and it was theoretically not impossible that they should have excavated deep canyons. The matter, especially of glacier efficiency in erosion – a vexed question – is one mainly of the evidences. It cannot safely be handled deductively; and it need not be, since in glaciated mountains the evidences crowd the field. But the announcement was unscientific, because unsupported by facts of observation. Its author had no right to make it. On the other hand, it was no less unwarrantable and dogmatic to assert the contrary, which also was freely done.

My own acquaintance with the phenomena of glaciation of the alpine type had its beginning in the Sierra Nevada, in 1883, in the latitude of the Yosemite Valley – the so-called High Sierra. Prevailing opinion as to that region, it appeared, ranged between the two extreme views indicated; namely that, as regards quantitative effects in degradation more especially, glaciation had been widely destructive of the preglacial topography, on the one hand; on the other, that it had been relatively protective. But there was no recognition of distinctive forms – beyond 'U-canyons' and moraines. I had little notion, therefore, as to what I should discover; only an open mind and a lively curiosity.

I was a maker of topographic maps, of some experience, and had a topographer's familiarity with the erosion aspects of mountains; but only of unglaciated mountains. I had as well, however, something of the inquisitiveness of the physiographer as to the origin and development of topographic forms.

The first station occupied in this work of survey was Mount Lyell, one of the most widely commanding summits of the vast mountainous tract of the High Sierra.

From Lyell there was disclosed a scheme of degradation for which I had not been in the least prepared. No accepted theory of erosion, glacial or other, explained either its ground-plan outlines or its canyon-valley profiles; and, so far as I can see, none makes intelligible its distinctive features now. The canyons, at their heads, were abnormally deep; they were broadly flat-bottomed rather than U-formed, the ratio of bottom width to depth often being several to one; and their head walls, as a rule, stood as nearly upright, apparently, as scaling of the rock would permit. I characterized them, figuratively, as 'down at the heel'. In many instances the basin floor, of naked, sound rock in large part, and showing a glistening polish on wet surfaces, was virtually without grade, its drainage an assemblage of shallow pools in disorderly connection; and not infrequently the grade was backward, a half-moon lake lying visibly deep against the curving talus of the head wall, and visibly shallowing forward upon the bare rock-floor.

The amphitheater bottom terminated forward in either a cross-cliff or a cascade stairway, descending, between high walls, to yet another flat. In this manner, in steps from flat to flat, commonly enough to be characteristic, the canyon made descent. In height, however, the initial cross-cliff at the head dominated all. The tread of the steps in the long stairway, as far as the eye could follow, greatly lengthened in down-canyon order. In that order, also the phenomena of the faintly reversed grade and of the rock-basin lakes rapidly failed. Apparently, at the canyon head, the last touch of vanishing glaciation had been so recent that filling had not been initiated, while downstream, incision of the step cliffs and aggradation of the flats had made at least a beginning in the immense task of grade adjustment; the tread of the step was graded forward, but so insensibly, as a rule, that its draining stream lingered in meanders on a strip of meadow, as though approaching base-level. These deep-sunk ribbon meadows, still thousands of feet above the sea and miles in length, reflecting in placid waters their bordering walls or abnormally steep slopes, presented an anomaly of the longitudinal profile in erosion no less impressive than that of the upright canyon heads.

In ground plan, the canyon heads crowded upon the summit upland, frequently intersecting. They scalloped its borders, producing remnantal-table effects. In plan as in profile, the inset arcs of the amphitheaters were vigorously suggestive of basal sapping and recession. The summit upland – the preglacial upland beyond a doubt – was recognizable only in patches, long and narrow and irregular in plan, detached and variously disposed as to orientation, but always in sharp tabular relief and always scalloped. I likened it then, and by way of illustration I can best do so now, to the irregular remnants of a sheet of dough, on the biscuit board, after the biscuit tin has done its work.

In large part, apparently, a preglacial summit topography had been channeled away. By sapping at low levels, by retrogressive undercutting on the part of individual ice-streams at their amphitheater heads in opposing disorderly ranks, the old surface had been consumed, leaving sinking ridges, meandering dulled divides, low cols or passes, and passageways of transection pointing to piracy and to wide shiftings of the glacial drainage. There was not wanting a scattering of the more evanescent sharp forms of transition which the hypothesis would require, as thin arêtes, small isolated table caps, needle-pointed Matterhorn pyramids with incurving slopes, and subdued spires (in the massive granite tracts) with radiating spurs inclosing basin lakes. The broader areas of this deep erosion, where complex channeling seemingly had passed into the phase of confluent glaciation, presented a much less intelligible ground-plan pattern. In every case, however, there was still an approximately central draining canyon. It was the sprawling high-walled masses of the residual uplands that told a clear story.

The legitimate inference was that the suspension of glaciation had suspended as well a process which had threatened truncation of the range. It was obvious, postulating recession, that the canyons of this summit region were independent in their courses, and had developed independently, of the initial upland drainage plan. It was clear, from their grade profiles, that they were not stream-cut. The inference was not only legitimate, but necessitated, that, profoundly deep as they were, they were essentially of glacial excavation.

Here, then, were facts of observation in support of one of the two extreme views referred to at the outset. That, however, ice-streams had descended on the long western slope to the range foot, or even close to it, I subsequently found to be untrue.

The canyons of that flank I now regard as stream work below, as ice work above, and as the joint product of streams and of glaciers, alternately, in between. But wherever they are abnormally deep, I infer from the evidence of the floor profiles that they have been thus deepened by glaciers.

The range had not been domed over by a continuous ice-sheet; it had been glaciated rather against its upper slopes. The summit tracts, narrowed by flank attack, had remained bare, perhaps because wind-swept; the ridges and peaks of degradation continued emergent, because sharp. Obviously, though its period had been short, the action of the process had been relatively rapid; for in the shallow pre-glacial canyons of the broad foothill zone, in which lay the moraines of the outer glacial boundary (the relatively insignificant, coarse products of degradation), the normal processes of erosion had accomplished so little that, seemingly, their action there had been suspended.

The summation of the hypothesis was that retrogressive cutting in large part had carried away the uplands, along an approximately definite, and an approximately level, plane of attack. Deep canyons had resulted, indirectly, because recession, directed horizontally, had been directed into a rising grade. This action seemed distinct from that of abrasion. Abrasion accomplishes deepening vertically and directly. In the case of a 'continental' glacier upon a level plain, abrasion would be operative alone. But that process was not to be invoked in explanation of the scalloped, tabular forms of the High Sierra; these pointed only to basal sapping.

Basal sapping in its details, however, was unintelligible. It was not immediately apparent, at least, how a glacier, originating against a precipitous rock slope, and drawing away from that slope, could undercut and cause it to recede.

DESCENT OF A BERGSCHRUND

To return to the narrative form, one feature of the small ice-body lying deep in a great amphitheater opening northward from Mount Lyell – one of perhaps a dozen 'glacierets' of the High Sierra – seemed to offer the explanation.

Among the numerous crevasses or schrunds of several diverse systems sharply lining the snowy surface upon which I looked directly down as upon a map, one master opening, the *Bergschrund* of the Swiss mountaineers, paralleled the amphitheater wall, a little out upon the ice. In detail it was ragged and splintered, but its general effect was that of a symmetrical great arc. I had already in mind, vaguely formulated, the working hypothesis that the glacier makes the amphitheater, that it is not by accident that glaciated mountains, and such mountains only, abound in forms peculiarly favorable to heavy snow-drift accumulation. My instant surmise, therefore, was that this curving great schrund penetrated to the foot of the wall, or precipitous rock slope, and that a causal relation determined the coincidence in position of the line of deep crevassing and the line of the assumed basal undercutting.

So much of assumption, so plausibly grounded, rendered direct observation at this critical point on the glacier floor compellingly desirable; and, returning to camp for all hands and the pack ropes, the rather appalling task was in fact very easily accomplished.

The depth of descent was about one hundred and fifty feet. In the last twenty or thirty feet, rock replaced ice in the up-canyon wall. The schrund opened to the cliff foot. I cannot say that the floor there was of sound rock, or that it was level; but there was a floor to stand upon, and not a steeply inclined talus. It was somewhat cumbered with blocks, both of ice and of rock; and I was at the disadvantage, for close observation, of having to clamber over these, with a candle, in a dripping rain, but there seemed to be definitely presented a line of glacier base, removed from five to ten feet from the foot of what was here a literally vertical cliff.

The glacier side of the crevasse presented the more clearly defined wall. The rock face, though hard and undecayed, was much riven, its fracture planes outlining sharply angular masses in all stages of displacement and dislodgment. Several blocks were tipped forward and rested against the opposite wall of ice; others, quite removed across the gap, were incorporated in the glacier mass at its base. Icicles of great size, and stalagmitic masses, were abundant; the fallen blocks in large part were ice-sheeted; and open seams in the cliff face held films of this clear ice. Melting was everywhere in progress, and the films or thin plates in the seams were easily removable.

These thinning plates, especially, were demonstrative of alternate freezings and thawings, in short-time intervals, probably diurnal. Without, upon the cirque or amphitheater wall, above the glacier, such intervals would be seasonal. Thus, apparently, to generalize from observation at a single point, the arc of the bergschrund foot, and the coincident arc of the cirque-wall foot, is a narrow zone of relatively vigorous frost-weathering. The glacier is a cover, protective of the rock surface beneath it against changes of temperature. Probably the bed temperature does not fall below that of melting ice. Hence, if (in summer) the bed at the wall foot is exposed, through the open bergschrund, to daily temperature changes across the freezing-point, frost-weathering must be sharply localized. The glacier will be efficient as the agent for débris removal; the result, therefore, must be quarrying and excavation, and basal sapping.

The amphitheater floor has been described as characteristically reversed in grade, though at a slight angle, and ponded backward against its head wall. It may be assumed that the disrupting action of frost at the bergschrund foot is directed against the floor, as well as against the cliff. Likely enough, also, the glacier is a highly efficient agent for removal, supplementing, by 'plucking', the initial rupturing work of frost. Its plucking action may be directed downward as well as backward; but downward action, at an early stage, will be defeated by the rapidly increasing difficulty of waste removal. The great arc-form crevasse at the glacier head, therefore, may be the indirect cause, not merely of recession of the canyon head at a low grade into a high grade, but of recession at a grade declining from the horizontal. Apparently there is a limit to such extension – a limit the earlier reached, the sharper the acclivity of the upland surface into which channeling is extended; and the head wall, in consequence, breaks back into steps, successively shortening in length of tread. The rearward steps may continue to be marked by schrunds rising to the glacier surface; living glaciers, in fact, are often characterized by 'cascades' in their upper courses; but sharply defined cross-cliffs, in empty glacial canyons, are rarely to be found far down-stream. Presumably they are so deeply buried there as to be wholly left to the dulling influence of scour. The 'rock-basin' lakes of the bare floors toward the head have received more attention than the great bowl of the amphitheater itself, the most phenomenal of all constantly recurring mountain forms; but I suspect that they are even more significant than has been supposed.

The hypothesis as to the action at the bottom of the bergschrund, in explanation of sapping, is slenderly supported. Physiographic inquiry has been an avocation merely, and I have failed of opportunity to repeat and to extend my early observations. Those observations, furthermore, were hastily and somewhat carelessly made, and were not recorded at the time. But that deep basal sapping in massive rocks especially, as in the summit region of the High Sierra, accounts for the anomalies of towering upland remnants, of canyons gradeless for miles, and of the sharp scalloping in ground-plan everywhere to be observed, is at once apparent, I think, upon full recognition of the forms themselves.

GLACIAL SCOUR AND VALLEY PROFILE

With these destructional effects assigned to glacial agency, a novel possibility is at once suggested as to the part played in their persistent development by glacial scour, or coarse abrasion. The upright element in the profiles, it would seem, must be regarded as a sapping effect in which scour plays no part at all. But the approximately horizontal element, considering its great extension, often, and the relatively abrupt descent by which compensation is made, constitutes a difficulty no less. The adjusted grade in river erosion is a smooth curve, lessening in declivity in the direction of flow. The glacier, however, by ablation, is diminished in volume as it lengthens; it is normally deepest close to its head; and possibly it is most effective in scour-erosion in proportion as it is deep. It must, in that event, tend to produce a valley 'down at the heel'.

The reverse grade, on amphitheater floors especially, occurs with sufficient frequency to be regarded as a type form. Rock-basin lakes, beginning at the amphitheater head, sometimes have notable length, several times the canyon width. The upper surface of the glacier here, on the other hand, invariably declines forward. Thus, in specific instances, it is not merely inference, but fact, that the glacier is deepest at the rear, and excavates there to a forward-rising grade.

It is, furthermore, implied that forward inclination of bed is not essential to glacier movement. It is not necessary, merely to determine that question, to inquire intimately into the nature of glacial motion. Fundamental in that motion, apparently, is the weight of the ice; and if the glacier at bottom, under its own weight, is not strictly viscous, it is apparently at least viscoid, responding in effect to the law of liquid pressures.

A viscous substance, heaped upon a level surface, spreads in mounded disk form, deepest at the center. Its flow-curve, in any radial vertical plane, advances from the bottom. The tendency to flow movement is proportioned to depth – to load; it diminishes toward the outer margin. The outer portions, therefore, move too slowly, and are affected by horizontal, forward thrust. They are retarded at the same time by basal friction, and in consequence present a bulged and swelling front, implying, over a broad marginal tract, rising lines of flow. But the glacier is terminated forward, and is thinned toward its termination, by combined melting and evaporation – i.e., by ablation; and, by ablation, it may be inferred, the constantly bulging front is planed away. The glacier may be regarded as made up of two layers – a superficial, relatively rigid layer, and a basal layer, mobile under the weight of the other; or of a zone of fracture and a zone of flow. In the thinning frontal region, the upper layer, or cover, is brought into contact with the bed. Rearward, it is lifted; though at the same time there it is planed away. Hence, rising lines of flow in effect extend to the surface; for the cover is to be regarded as a zone of rigidity merely, constant only as to position, and thickening, from the mobile ice below, as it is thinned by ablation above. Rates of glacial motion, measured along the surface, therefore will be deceptive. On these assumptions, the line of most rapid advance in the glacier mass is from near the bed, at the rear, to the surface, near the front. Along the bed, motion slows forward; and as pressure upon the bed diminishes in that direction, presumably abrasive erosion is most vigorous toward the rear. The accepted view as to the flow-curve of the river is that, normally, it advances most rapidly at the surface. Deep rivers, however, are found to advance from a point measurably below the surface. If rivers had the great depth of glacial streams, possibly it would appear that the curve of flow which they actually have is but the reverse curve due to bed friction, extended to the surface because the surface is near. It would seem to be a safe assertion that descending grade of bed is not essential to river motion, only decline of the river surface toward the level of discharge; and that, in a long canyon with level floor, terminating at the sea, a river, one or two thousand feet in depth and maintained at that depth at its head, would advance with essentially the same flow-curve as that here attributed to the glacier. The value of such speculation consists in the indication it affords that appeal to the observed flow-curve of the river, in rebuttal, may not be valid.

The long ribbon meadow of the lower canyon course, no less than the ponded amphitheater floor, I think, invites interpretation as the manifestation of a tendency on the part of the glacier to channel excessively up-stream. And in this overdeepening toward the canyon head, I suspect, the two agencies of horizontal sapping and of vertical corrosion powerfully co-operate.

The ultimate effect, upon a range of high-altitude glaciation, would be rude truncation. The crest would be channeled away, down to what might be termed the base-level of glacial generation. Where, among the determining causes of glaciation, high latitude rather than high altitude is operative, the base-level of degradation may lie below the sea, deepest centrally and shallowing outward. Given a land area initially, the glacier itself, as degradation approached its maximum, would replace the land, affording the necessary above-sea surface for snow accumulation. The degradation limit would be determined by the lifting power of the sea.

The hypothesis, at this stage, is of much less importance than recognition of the anomalies of fact, of which it offers a tentative, even venturesome, explanation. In the fiorded regions of the globe, notably in the Patagonian Andes, of which a well-controlled reconnaissance survey has recently been completed, we have examples not only of fiords deeping backward for many miles into rising grades, but of fiord lakes, in parallel series, penetrating from foot-hills on the one side to foothills on the other, transecting a range. In explanation of such deep channels, whether occupied by arms of the sea, by lakes, or by feebly moving streams on meander bottoms, the appeal to grades, it seems to me, will be most cogent.

4 Crescentic Gouges on Glaciated Surfaces

G. K. GILBERT

ASSOCIATED with striae and other evidences of glacial abrasion are certain types of rock fracture. These have been classified and described by T. C. Chamberlin (1888) in his 'Rock-scorings of the great ice invasions'. The three principal types are chatter-marks, crescentic cracks, and crescentic gouges. All of these are so associated with glacial sculpture and striation as to indicate that they are of glacial origin. They all occur characteristically in sets, the members of each set succeeding one another in the direction of ice motion and each individual marking having its longer axis athwart the direction of ice motion.

Chatter-marks and crescentic gouges have a common character, in that each is characteristically a shallow furrow with crescentic outline. In crescentic gouges the convexity of the crescent is usually turned forward in the direction of ice motion; in chatter-marks it is usually turned backward. Chatter-marks are closely associated with grooves engraved by boulders. Crescentic gouges are not thus associated; they frequently occur on surfaces exhibiting no other marks of glaciation except fine striae and polish.

The features called crescentic cracks are vertical fractures of the rock without the removal of fragments. They are usually curved in plan, with the concavity turned forward. Their orientation thus relates them to the chatter-marks, but they are independent of grooves.

An approximate understanding of the glacial chatter-mark is easily reached, because the phenomenon is intimately related to the chatter-mark of the machinist, from which it is named. The plowing of a groove in a brittle substance is not a continuous process, but is accomplished by making a series of fractures, each one of which separates a fragment of the substance. Each fracture is preceded by a condition of strain and stress, and these are relieved by the fracture. The resistance to the grooving tool is thus essentially rhythmic, and if the tool is slender, or is not firmly supported, a vibratory motion is set up (with chattering sound) and the groove becomes a succession of deep scars. When the grooving tool is a hard boulder held in a slow-moving body of ice, and the thing grooved is a brittle rock, the remaining condition for rhythmic action is probably found in the elasticity of ice and rock, which permits the development of strain and stress before each fracture.

The crescentic crack, being vertical, is presumably a result of tensile stress parallel to the rock face. As the glacier moves forward it tends, through friction, to carry the bed-rock along with it. If the friction on some spot is greater than on the surrounding area, the rock just beneath that spot is moved forward in relation to the surrounding rock through a minute but finite space. This relative movement involves compression about the down-stream side of the affected rock and tension about its up-stream side, the magnitude of the stresses depending on the differential friction, and rupture ensuing when the tensile stress exceeds the strength of the rock. Exceptional friction may be given by the passage in the ice base of some substance which has a high coefficient of friction in relation to the bed-rock; for example, if the glacier base contains a pocket of sand surrounded by clear ice, the coefficient of friction between the sand and the bed-rock will probably be much higher than between the ice and the bed-rock.

The crescentic gouge is less easy to understand, and it is the purpose of this communication to put forward a hypothetic explanation.

Crescentic gouges have been observed in granite and other massive plutonic rocks, in sandstone, and in limestone. My own observations have been made chiefly in the granite district of the High Sierra, where opportunities for the study of glacial sculpture are exceptionally good. In some localities the gouges are abundant, and in most districts where glacial polish and striation are preserved they can be found by a few minutes' search. Since my attention was specially directed to the question of their origin I have examined several hundred.

In length they measure from a few inches to more than 6 ft, measurement being made in a straight line from horn to horn of the crescent. Within the range of my observation chatter-marks are comparatively small, the largest observed being less than a foot in length. Solitary gouges are often seen, but in the majority of cases they occur in sets of from two to six or seven. Ordinarily the members of the same set have about the same size, but in a few cases a progressive increase was observed, the individual most advanced in the direction of ice motion being largest. It seems legitimate to infer from this arrangement in sets that the cause of the gouge, whatever it may be, moved forward with the ice. As already mentioned, the convexity of the crescent is turned forward: but to this rule there are occasional exceptions. Two or three individuals were seen with the concavity turned forward, and a few also with the longer axis in the direction of ice motion. The gouges were seen only on the upstream sides of projecting bosses (Fig. 4.1). They are not restricted to the bottoms of glacial troughs,

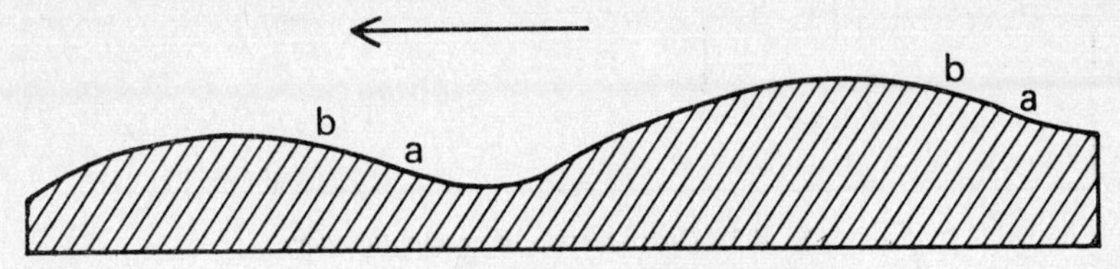

Fig. 4.1 Profile of part of glacier bed

The arrow shows the direction of ice movement. Crescentic gouges occur on ascending slopes, from *a* to *b*.

but occur also on the walls, and in that case are on the upstream faces of salients.

CONOID FRACTURE

The cross-profile of the crescentic gouge (Fig. 4.3) exhibits an angular notch bounded by two unequal slopes. The slope from the upstream edge is gentle, that from the downstream edge approximately vertical. This character is exhibited in all parts of the crescent (Fig. 4.2). The gentler slope radiates from an axis somewhere within the curve of the

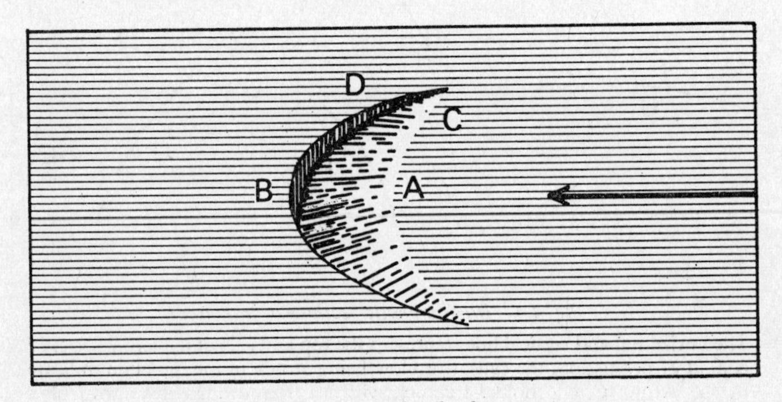

Fig. 4.2 Diagrammatic sketch of crescentic gouge

The arrow shows the direction of ice movement. Compare Fig. 4.3.

crescent and is essentially a portion of a conoid surface. It is one wall of a fracture, or crack, which does not end at the bottom of the gouge, but continues on into the rock to an undetermined distance. The fact that the vertical fracture terminates against the oblique fracture shows that the oblique was first made. The oblique, or conoid, fracture may therefore be regarded as the primary product of the causative force,

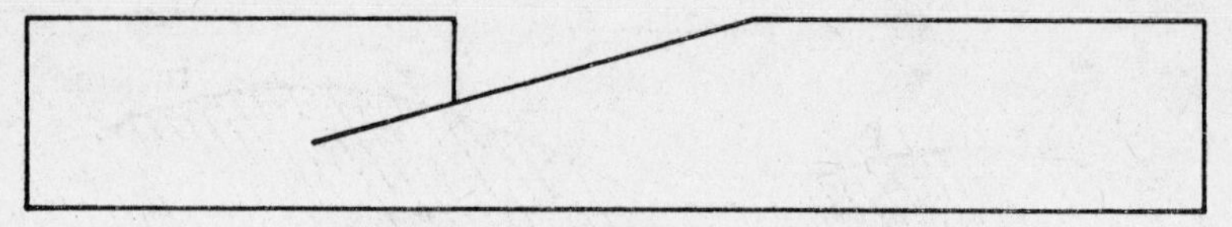

Fig. 4.3 Cross-section of crescentic gouge
The section represented is from *A* to *B* or *C* to *D* of Fig. 4.2.

and the vertical fracture as secondary; and in seeking a cause of the phenomenon I have given first attention to forces which might be appealed to in explanation of the conoid fracture.

There is another conoid fracture with which geologists are familiar, the fracture often made in obsidian, or other homogeneous brittle rock, by a light blow of the hammer. This is sometimes called the conoid of percussion (Fig. 4.4). Usually it circles completely about its axis, but sometimes it is one-sided. Its relation to the surface

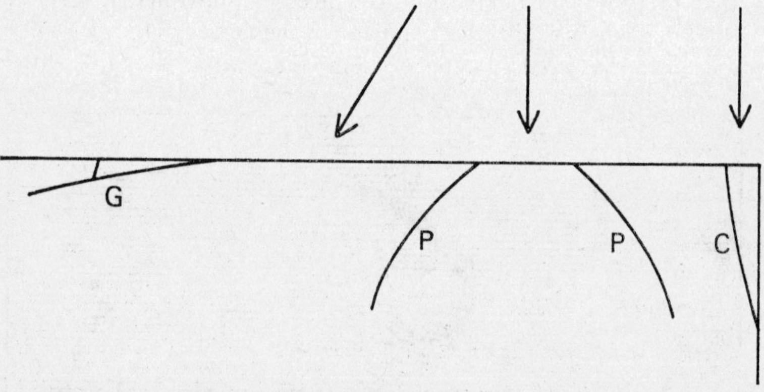

Fig. 4.4 Diagrammatic sections of fractures
The horizontal and vertical lines represent the top and side of a mass of rock; *PP*, conoid of percussion, the causative blow being struck in the direction indicated by the arrow; *C*, conchoidal fracture, with arrow showing direction of blow; *G*, conoid and vertical fractures of crescentic gouge, with oblique arrow showing theoretic direction of causative pressure.

struck by the hammer resembles closely the relation of the glacial conoid to the external surface of the bed-rock, and the one fracture may help to explain the other. The conoid of percussion is caused by a blow; that is, by the instantaneous application of pressure to a small area. No way has occurred to me in which a glacier can make such a fracture by means of a blow, but it seems possible, as I shall presently explain, that a glacier can slowly apply considerable differential pressure to a very restricted area; and there is some reason to believe that suddenness of impact is not essential to the production of conoid fractures. The ordinary conchoidal fracture, which is a near relative of the conoid of percussion, is commonly developed in brittle rock by a blow struck near an edge (Fig. 4.4); but it is also produced by simple pressure in the manufacture by Indians of flint and obsidian implements.

In a single instance a conoid fracture in granite was observed to circle completely around its axis, thus simulating still more closely the conoid of percussion. This feature occurs in a glacial trough where there are many crescentic gouges, but it is not connected by gradation with the ordinary gouges.

DIFFERENTIAL PRESSURE

As a glacier moves forward its under surface is continuously adjusted to the irregular shapes of its bed. The greater inequalities of the channel find expression on the upper surface of the glacier, but the minor inequalities do not affect the upper part of the ice stream. The diagram (Fig. 4.5) represents in profile a projection of the bed, of

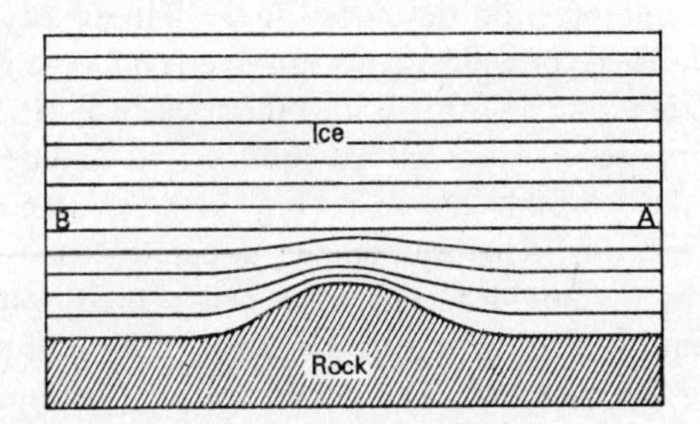

Fig. 4.5 Longitudinal section of lower part of a glacier

The section is supposed to be at a point where it passes a projection of the rock bed, and illustrates the deflection of lines of flow and the temporary compression of the lowest layers of ice.

moderate magnitude in relation to the total thickness of the glacier. The adjustment of the glacier to this obstruction affects the flow lines of only the lower strata of ice, leaving unaffected all above some limiting plane, *AB*. Below that limit the lines of motion first ascend and then descend in passing the obstruction. If we think of the flow lines of the diagram as separating layers of ice, then each layer becomes thinner in ascending and gains thickness in descending.

A large boulder embedded in the glacier close to its base is not reduced in thickness along with the enclosing ice layers, and its

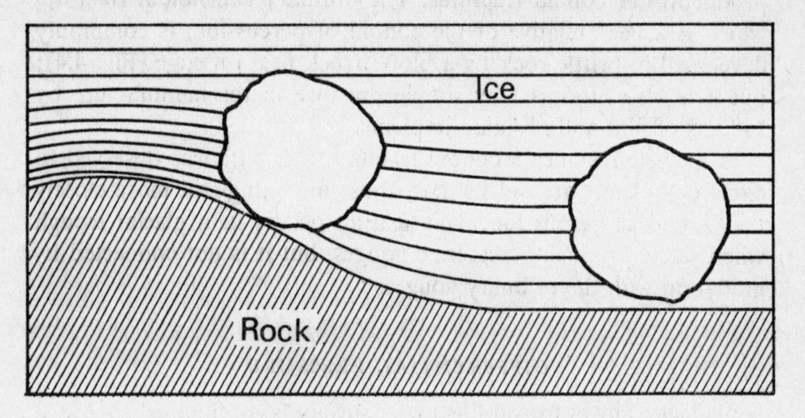

Fig. 4.6 Longitudinal section of lower part of a glacier

The section is on the upstream side of a projection of the rock bed and illustrates the changing relations of an embedded boulder to the system of flow lines. The ice motion is from right to left.

resistance to compression develops differential stresses both above and below it. These pressures tend to force it into the overlying ice, and at the same time to force it into the rock bed. As the rock bed effectually opposes the downward motion the boulder is actually forced into the ice body above it (Fig. 4.6). A large share of the pressure thus brought to bear on the upper side of the boulder is transmitted by the boulder to the rock bed at their point of contact or approximation, and there is thus a concentration of pressure on a small area of the rock bed. This concentration continues as long as the ice about the boulder is undergoing vertical compression in passing the projection, and ceases when that compression ceases. It begins somewhere on the upstream slope of the projection and ceases at its crest. Thus the conditions for localization of pressure by this method have the same distribution in relation to prominences of the rock bed as that observed for the crescentic gouges.

If the ice beneath the boulder is clear of débris, it is probable that a large differential pressure cannot be developed without causing the ice to flow away and bringing the boulder into contact with the rock bed. The result to the rock bed of such contact would be a deep scratch or groove, and grooves are not the normal associates of crescentic gouges. It seems necessary, therefore, to suppose that when crescentic gouges were made the direct contact of the boulder was in some way prevented – and the means of prevention is not far to seek. If only the sand and other fine detritus normally abundant in the base of a glacier be assumed to saturate the ice beneath the boulder, a cushion is provided quite competent to prevent actual contact of boulder and rock bed and at the same time transmit the pressure of the boulder to a small area of the bed.

A complete discussion of this hypothesis would include a mathematical analysis of the mechanics of the conoid fracture. Only the elastician is competent to make such an analysis, and I have not attempted it. Nevertheless, as I have not been able to ignore altogether that aspect of the subject, I shall venture a few lay suggestions.

As the conoid of percussion is symmetric about an axis normal to the surface receiving the blow, and as the conoid of the crescentic gouge is asymmetric, it may be inferred that the direction of the force producing the latter is oblique to the rock surface. In a general way all pressures of the ice upon glaciated bed-rock must be oblique; otherwise there would be no forward motion; but the particular pressure to which appeal has been made in connection with the crescentic gouge is the result of a compression of the ice in the direction normal to the rock face, and should be regarded, I think, as itself normal. If this view is correct, some other cause must be sought for a special stress component parallel to the rock face.

I think such a cause exists in differential friction. The friction per unit area of the glacier on its rock bed at any point is the measure of the force there applied by the glacier in a direction parallel to the local rock surface. It varies with the material of the two bodies in contact and is directly proportional to the force, normal to the contact surface, by which they are pressed together. Therefore during the period in which the hypothetic boulder communicates an excess of pressure to a small area of the rock bed, the same area experiences a proportionate excess of sliding friction, and is consequently subject to a proportionate excess of force in a direction lying in the plane of contact. The composition of this force with the differential force normal to the plane of contact gives a resultant parallel to the general system of oblique stresses in the surrounding ice. This reasoning appears to warrant the statement that the differential pressure occasioned by the approach of the boulder to the prominence of the rock bed is oblique to the local rock surface and is directed forward.

DEFORMATION AND RUPTURE

To obtain an idea of the nature of the deformation resulting from the differential pressure just mentioned I have tried a few simple experiments. If a liquid jelly be allowed to cool in a large bowl it assumes

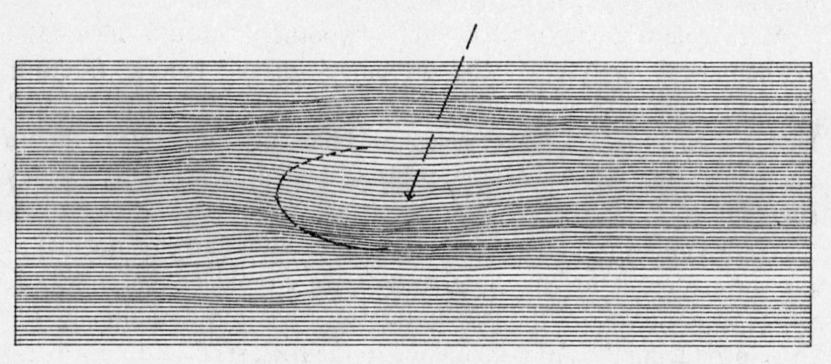

Fig. 4.7 Ideal oblique view of originally plane rock surface
Showing, with vertical exaggeration, its theoretic deformation under the external stress causing a crescentic gouge. The arrow shows the direction of the stress. The direction of ice motion is from right to left. The position of the conoid fracture is indicated by a broken line. Compare Fig. 4.8. The lines of the drawing are parallel equidistant profiles of the deformed surface.

the condition of an elastic solid with a level and smooth upper surface. Pressed by the ball of the finger its surface is deformed, the hollow under the finger being surrounded by a low circling ridge, the slope of which is relatively steep toward the finger but very gentle in the opposite direction. If the pressure of the finger be made oblique the ridge becomes steeper and higher on one side of the hollow, and is correspondingly reduced or even abolished on the opposite side. The hollow under the finger is a direct result of the pressure and the curving ridge is an indirect result, the intermediate factors being a complex system of internal strains and stresses.

I conceive that an analogous condition obtains in the rock bed as a result of the oblique pressure under the hypothetic boulder; that there is a central depression of the surface (Fig. 4.7); that this is margined on one side by a curved elevation; and that there are internal strains and stresses; but the strains are comparatively small and the slopes of deformation are very gentle, because in rock the strain limit is quickly reached and rupture ensues. The hypothesis

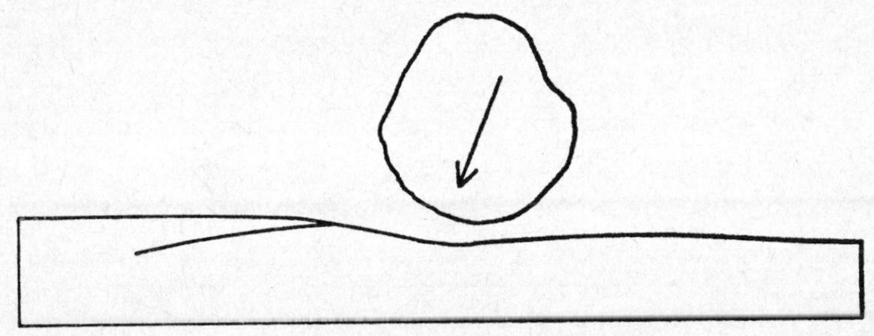

Fig. 4.8 Theoretic deformation of rock beneath a glacier

This ideal section illustrates the theoretic deformation of rock beneath a glacier by differential pressure in connection with an embedded boulder. The arrow indicates the direction of the pressure. The direction of ice motion is from right to left. The conoid fracture of the crescentic gouge is shown at left of the boulder. Compare Fig. 4.7.

assumes that rupture in this case is initiated in the surface of the rock, along the inner slope of the curving ridge (Fig. 4.7), and is propagated obliquely downward, forming the conoid fracture (Fig. 4.8) of the crescentic gouge.

In the absence of a rigorous analysis of the stresses associated with the deformation, the correlation of the conoid fracture with the curved ridge is an assumption only; but having made that assumption it seems possible to base on it certain inferences tending to throw light on other elements of the gouge. In the production of the deformation the rock compressed vertically under the boulder experienced horizontal dilatation whereby the ridge was pushed up, and the ridge itself experienced horizontal compression. The region of the fracture was thus subjected to horizontal compression just before the rupture, and as soon as the fracture had been formed the wedge of rock above it was relieved of horizontal compression and recovered its original horizontal extent. The wedge had also been bent, its upper surface constituting the crest of the ridge, and when it was detached beneath it tended to recover also its original unbent form by lifting its edge. This change was resisted by the pressure of the overlying ice, with the result that the wedge became affected by the stresses of a bent beam, compressive below and tensile above. To the tensile stress along the upper part of the wedge I ascribe the vertical fracture which completed the gouge. It is possible that more

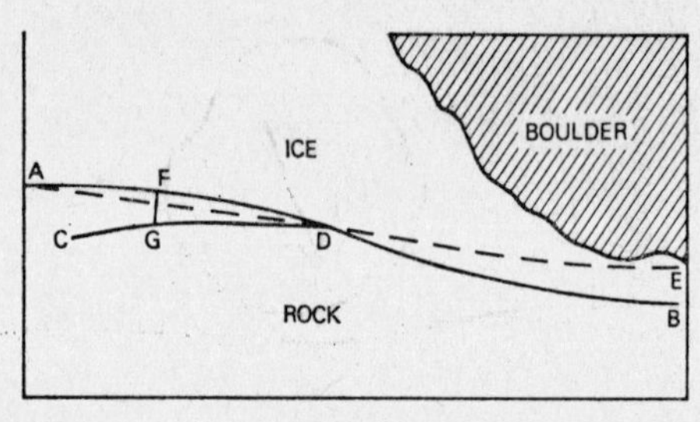

Fig. 4.9 Theoretic origin of fractures producing the crescentic gouge

This ideal section illustrates the theory of origin of fractures producing the crescentic gouge. *AE*, original (longitudinal) profile of rock bed; *AFDB*, deformed profile of rock bed (with exaggeration of curvature); *DGC*, conoid fracture; *FG*, vertical fracture.

than one vertical fracture ordinarily occurred, dividing the wedge into several parts.

The position of the vertical fracture, as thus explained, is conditioned (in part) by the distance to which the conoid crack penetrates the rock. Toward the horns of the crescent, where the conoid crack vanishes, the crack probably penetrates less deeply, and the vertical fracture there traverses a thinner part of the wedge; hence the curve given by the intersection of the vertical fracture with the surface is not concentric with the corresponding curve for the conoid fracture, but meets it. Some of these relations are diagrammatically shown in Fig. 4.2.

It is worthy of note that the two fractures, referred respectively to shearing and tensile stresses, differ notably in the textures of their surfaces. The conoid fracture gives a rather smooth surface, and that produced by the vertical fracture is comparatively ragged. The vertical fracture also departs from regularity more widely than the conoid.

Since the writing of this paper it has been examined by several friends well qualified to discuss its theoretic part, and as a result of various suggestions and criticisms a number of passages were modified. One of the most important suggestions – but one which I was unable to accept – included the following alternative hypothesis for the origin of the two fractures delimiting the crescentic gouge. I

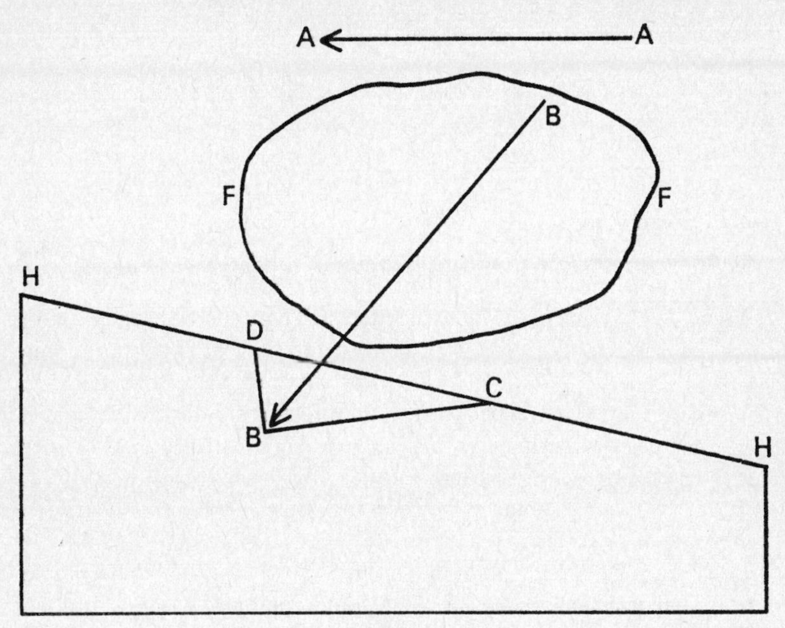

Fig. 4.10 Theoretic origin of the oblique and vertical fractures of the crescentic gouge

This diagram illustrates the alternative hypothesis for the origin of the oblique and vertical fractures of the crescentic gouge.

quote: 'We will suppose the glacier is moving in the direction indicated by the arrow *AA*; that the rock floor is indicated by *HH*; that the boulder *FF* is either in contact with or near the rock floor. Under these conditions the direction of the greatest stress would be indicated by the arrow *BB*, being the resultant of the weight of the glacier and the pressure behind the moving mass. Under these conditions there are powerful shearing stresses in the directions *BC* and *BD*. These stresses are greater adjacent to the boulder because it is a rigid body and is able to transmit forward close to the rock floor the pressure of the ice about it. At the place where there is the most rapid change in the amount of compression rupture takes place. Whether the rupture occurs in both the horizontal and vertical directions will depend upon circumstances, which will largely depend upon the shape of the rock surface and the position of the boulder and its shape. When the vertical rupture takes place alone you have the crescentic cracks, when horizontal rupture takes place, followed instantaneously by the vertical rupture, you have the crescentic gouges.'

RHYTHM

A full development of the hypothesis would include also a discussion of the occurrence of the gouges in series, and this likewise requires the expert knowledge of the elastician. If I again venture a suggestion it is largely in the hope of exciting his interest. There can be little question that each series of gouges represents a mechanical rhythm of some sort. In a large group of mechanical rhythms, including many in which friction plays a part, a force uniformly applied accumulates strain and stress, which are relieved in some catastrophic manner whenever they reach a certain limit. In the present case the conoid rupture is a catastrophic event relieving some of the internal stresses of the bed-rock. The jar, or miniature earthquake, occasioned by it and radiating from the point of rupture may be supposed to overcome frictional resistance between glacier and rock and cause a sudden slipping along their contact surface, thus relieving the frictional strains and stresses for some distance in all directions. The boulder instantaneously moves forward to a new position with reference to the rock bed, and the gradual renewal of deformation and internal strains is begun. This line of inference leads to the difficult question whether the sudden forward movement covers only the fraction of an inch, or whether it may be of the order of magnitude of the interval between gouges – from a few inches to several feet. If it is very small, then the determination of the gouge interval remains as one of the obscure factors of the hypothesis. In a general way the gouge interval is related to the gouge length, being greater when the length is greater, but there is no fixed ratio between the two. No measurements were made in the field, but photographs show a range in the ratio from about 1:3 to about 2:1.

The discussion of the gouge rhythm also suggests the possibility that the ordinary movement of the glacier on its bed may be rhythmic. It is certainly conceivable that internal strains and stresses of the rock and ice up to the limit given by static friction may be locally engendered during periods of adhesion and then relieved by momentary slipping, with sliding friction only.

RESISTANCE OF ICE TO FLOWAGE

The crescentic gouge is a large disruptive scar on the face of a compact jointless rock. Any hypothesis to account for it must provide great force. The particular hypothesis here given appeals to the differential stress developed by the resistance of the ice to the forcing of a boulder into it. It can not be true unless the ice has great power of resistance to flowage; and, conversely, if it is true, the ice has greater power of resistance than some students have been disposed to admit. It is generally understood that cold ice is more rigid than ice at the melting temperature, but the hypothesis is not concerned with cold ice. Doubtless crescentic gouges are made under cold ice, but the gouges preserved for our observation were not so made. Beneath the forward part of a glacier the basal temperature is the temperature of melting (as conditioned by the pressure); and as a great glacier wanes, every portion of the bed is in turn subject to the action of its forward part. The finishing touches, therefore, the surface markings and the small details of sculpture, can not be ascribed to ice of the low temperatures theoretically obtaining far back under the névé. So the crescentic gouges, as explained, testify to the resisting power of ice at the most favorable temperature for flowage.

Whether we regard ice as a plastic substance, or whether we accept, as I do, the view of Chamberlin, that it is made up of rigid crystalline grains and flows chiefly by interstitial melting and regelation, we must recognize a relation between velocity and resistance to flow. The more rapid the flow the stronger the resistance. Therefore the crescentic gouges, if they have been properly explained, may testify also to relative rapidity of glacier movement.

5 Geologic History of the Yosemite Valley

FRANÇOIS E. MATTHES

IN no part of the Sierra Nevada have the evidences of glacial action been studied in greater detail than in the Yosemite region. John Muir was the first to engage in this work. During the years of his residence in the Yosemite Valley he devoted a large part of his leisure time to tracing the pathways followed by the ancient glaciers, not only in the immediate vicinity of the valley but also in the High Sierra above. Thirty years later Henry W. Turner, of the United States Geological Survey, undertook to locate the larger glacial deposits on a small-scale map, but he did not finish the task. It was not until 1913 that a systematic and detailed glacial survey – the survey on which the present account is based – was instituted and that the data obtained were assembled on a large-scale map. In subsequent years this survey was extended to the High Sierra as well as to the country below the Yosemite region, until at length the entire area that once lay under the dominion of the Yosemite Glacier, with its tributaries and neighbors, was covered. As a result, the length, breadth and depth attained by each of these glaciers at the time of their greatest extension are now definitely known. What is more, some insight has been gained into the history of their advances and recessions in response to climatic changes, so that it is now reasonably certain that the Yosemite Valley was invaded by a glacier three times. There is even some warrant for the presumption that the valley was invaded more than three times, the glacial history of the Sierra Nevada, like that of the northern parts of the continent, having consisted of several wintry subepochs, or 'glacial stages', alternating with mild subepochs, or 'interglacial stages'.

EVIDENCES OF GLACIAL ACTION

The evidences of glacial action in a mountain region are in general of three kinds – grooved and polished rock surfaces, characteristic topographic forms, and deposits of ice-borne rock débris. Of these three kinds of evidence the first would naturally suggest itself, because of its highly distinctive nature, as the easiest to recognize and therefore the most valuable for the identification of ancient glacier paths.

Glaciers, indeed, literally grind, tool and polish their beds with the rock débris that is frozen in their bottom layers. The large angular blocks produce grooves a fraction of an inch in depth; the smaller fragments produce fine parallel 'striae'; and the sand and mud scour and polish the rock until it fairly gleams. The resulting striated and glassy rock surfaces are familiar to all who have visited intensely glaciated mountain areas. In the upper Yosemite region and the adjoining parts of the High Sierra such surfaces are particularly plentiful and extensive; there one may walk on 'glacier polish' for considerable distances.

However, these vivid evidences of glacial action are not as a rule long lived. The reason is obvious; the rock, being constantly exposed to the weather and the wearing action of running water, inevitably decomposes and disintegrates in the course of time; the polish flakes off, the surface becomes increasingly rough and irregular, and the evidence of glaciation disappears. Some types of rock, naturally, weather more rapidly and lose their polish sooner than others, and this is true also in the Sierra Nevada. To assume, therefore, that the ancient glaciers on that range extended no farther than the polished and striated rock surfaces seem to indicate would be clearly unjustifiable. Actually the glaciers extended much farther; they spread over a vast contiguous territory from which all vestiges of glacial abrasion have now vanished.

Neither can the testimony of the topographic forms be safely relied upon. Among the more conspicuous forms that are generally held to be characteristic products of glaciation are U-shaped, trough-like canyons having spurless, parallel walls and stairwise descending floors that commonly hold shallow lake basins on their treads; also, hanging side valleys from whose mouths the waters cascade abruptly down into the main canyons. Such forms the reader already knows to be abundantly represented in the Yosemite region, and it might therefore seem to him that it would be a simple matter to determine the limits reached by the ice by observing how far these telltale forms extend down the flank of the range. However, it is to be borne in mind that the development of characteristic glacial forms is conditioned in part by the nature of the rock. Glaciers do not work with equal facility in all kinds of rock. As will be shown more fully further on, they work to best advantage in soft rocks and closely jointed rocks. In massive or sparsely jointed granite they accomplish relatively small results and seldom produce characteristic glacial forms. Thus it is that in the Sierra Nevada, where granitic rocks of massive habit prevail over large areas, many canyons and valleys retain in large measure their preglacial V shapes in spite of repeated vigorous glaciation. A conspicuous example is the Grand Canyon of the Tuolumne River, which lacks for the most part the characteristic U form of glaciation, although it was the pathway of the greatest and most powerful ice stream in the Sierra Nevada. Again, some valleys and canyons, though considerably remodeled by the ice, lack steps and lake basins in their floors, as, for instance, the valley of the Dana Fork of the Tuolumne River.

Side valleys can be left hanging as a result of other than glacial processes, notably by the rapid trenching of a master stream whose course has been steepened, in consequence of tilting, more than the courses of its tributaries. The Sierra Nevada abounds in hanging valleys of just such an origin, the Merced and the other southwestward-flowing master streams having intrenched themselves rapidly in consequence of the tilting of the range, whereas many of the small tributaries trending at right angles to their courses have not been able to trench at the same rate. In the Sierra Nevada, therefore, hanging valleys cannot be accepted as prima facie evidence of glacial action.

The most reliable record of glacial activity, on the whole, is that embodied in the deposits of rock waste left behind by the glaciers, particularly those ridge-shaped deposits termed 'moraines'. These unobtrusive features, wherever well preserved, accurately define the limits reached by the ancient glaciers. It was, therefore, with the moraines that the survey was mainly concerned. In the glacial survey of the Yosemite region all identifiable moraines and remnants of moraines, even isolated glacial boulders, were duly located and mapped, and thus there is now at hand a detailed record of the successive advances and recessions of the ancient glaciers.

TESTIMONY OF YOUNGER GLACIATED ROCK SURFACES

The floors and walls of the canyons that were the pathways of the later glaciers appear remarkably fresh and almost unweathered. Over surprisingly large areas they retain their polish and striae and are so smooth and glassy that walking or climbing over them with hob-nailed shoes is hazardous, and travel with horses or mules is impracticable. In many places, it is true, the polish has flaked off and the surface of the rock is rough, but the deeper scorings and flutings are still visible. Elsewhere plates of rock a quarter of an inch to perhaps a full inch in thickness have burst off or are in process of being loosened, but even there the smooth-flowing contours produced by glacial abrasion remain.

No one who visits the upper Yosemite region or the adjoining parts of the High Sierra can fail to be profoundly impressed by these facts. Few mountain regions, indeed, exhibit glacially worn and polished rock surfaces on a larger scale; few give the traveler a more vivid sense of the recency of the ice age. Scarcely credible does it seem to one viewing the vast expanses of gleaming granite that fully 20,000 years may have passed since the ice age came to an end; that even in the higher parts of the range, where the glaciers lingered long after they had receded from the canyons below, the rock has been exposed to the weather several thousand years.

Two circumstances explain the unusual abundance of glacier polish in the region above the Yosemite Valley – the prevalence of highly siliceous, slow-weathering types of granite, and the generally massive, sparsely jointed structure of those rocks. The superior durability of siliceous granite is strikingly demonstrated in many places where such granite is contiguous to a weaker rock, as, for instance, diorite. The granite as a rule still gleams with glacier polish, whereas the diorite has a roughened and perceptibly lower surface. Veins of hard, fine-grained aplite stand out in relief, little narrow causeways with level, polished tops, raised half an inch or more above the rough surface of the coarse granite, or granodiorite, which they transect. Doubtless the polish itself in such places helps to accentuate the difference, for it acts in some measure as a protective coating: it promotes the quick run-off of water from the surface, thereby lessening the proportion absorbed by the rock, and it retards the growth of lichens and mosses, thereby lessening the supply of carbonic acid and vegetable acids which result from the decay of those plants and which attack the weaker minerals.

The massive structure of the granite favors both the production and the preservation of glacier polish. Where the joint fractures are spaced far apart – tens or even hundreds of feet – the glaciers can not pluck or quarry individual blocks but work wholly by abrading, and the conditions are most propitious for the development of continuous expanses of even, polished rock; and there also few avenues are available through which water may penetrate to some depth below the surface, and thus the destructive action of such percolating water by hydration, solution, or freezing is reduced to a minimum.

That the glacier polish is by no means equally distributed throughout the Yosemite region and the adjoining High Sierra readily follows from the foregoing considerations. The distribution is controlled by three independent factors – the mineral composition of the rock, the joint structure, and the length of exposure since the retreat of the ice. Thus it happens that in the Yosemite Valley itself glacier polish is on the whole rather scarce, for the chasm was evacuated by the ice soon after the climax of the last glacial stage, possibly as long as 30,000 years ago; and, besides, its walls are made up in many places of jointed rock that was quarried rather than abraded by the glacier. As might be expected, the polish that remains occurs mostly on bodies of extremely durable and massive rock. Some of it, unfortunately, is inconspicuous or hidden from view and consequently is readily overlooked. A few small patches, remarkably well preserved, are on a buttress at the eastern base of El Capitan; other patches occur at the foot of the Three Brothers, at the base of the cliffs under Union Point, on the platform above the Lower Yosemite Fall, on the sides of the Washington Column, and on the walls near Mirror Lake. Glacial grooves remain visible, though the polish has disappeared, on the buttress west of the Royal Arches, on the cliffs east of Indian Canyon, and on the wall below Union Point. (See Fig. 5.12 for the location of some of these places.)

More abundant is the glacier polish on the platform above the Nevada Fall, both owing to the more recent glaciation of that platform and owing to the durability and massive structure of the Half Dome quartz monzonite, of which it is made. Indeed, the tourist as a rule catches the first glimpse of glacier polish when he arrives at the top of the Nevada Fall. The isolated patches he beholds there, however, measure but a few square yards each and are insignificant compared with the larger tracts to be seen on the floor and sides of the Little Yosemite and on the rounded backs of Liberty Cap and Mount Broderick. These tracts, in turn, seem small in comparison with the vast expanses of glacier polish that occur on the broad floor of massive granite of the upper Merced Canyon, above the Little Yosemite. The trail that leads to Merced Lake takes the traveler over this floor and affords him an excellent opportunity to view this unique area of burnished pavements and slopes.

Almost equally remarkable for its wealth of glacier polish, but far less accessible, is Tenaya Canyon. Indeed, the prevalence of glacier polish on its steeply sloping sides adds greatly to the difficulties which this extremely rugged canyon presents to those who would traverse it. Thus far only a handful of experienced and daring climbers have had the hardihood to pass through the entire length of the chasm.

TRANSFORMATION OF THE YOSEMITE VALLEY BY THE ICE

What, it may now be asked, was the effect of these repeated ice invasions upon the configuration of the Yosemite Valley? To what extent has the valley been remodeled by the glaciers, and in what measure is its present form due to their action? These are the questions that contain the kernel of the Yosemite problem.

As regards the capacity of glaciers to excavate deep canyons in hard rocks, little doubt now remains. Though there are still skeptics, it is safe to say that most geologists are convinced that glaciers are indeed powerful eroding agents. It is clear also that much depends upon the character of the rocks involved: in some kinds of rock, glaciers erode much more effectively than in others. Probably few glaciated valleys that have come under investigation recently have shed more light on this question than the Yosemite itself, for not only does this valley afford striking illustrations of the degree to which the efficiency of the glacial processes is influenced by the character and the structure of the rocks, but its preglacial depth and form have been determined with a fair degree of accuracy, and hence it is possible to calculate with corresponding accuracy the amount of rock material that was actually excavated from the valley during glacial time.

DEEPENING EFFECTED BY THE YOSEMITE GLACIER

The downward excavating effected by the Yosemite Glacier is clearly indicated by the longitudinal profiles in Fig. 5.1. All the space between the profile *C–C′* of the preglacial canyon stage and the profile *D–D′* of the glaciated rock floor of the valley (the bottom of

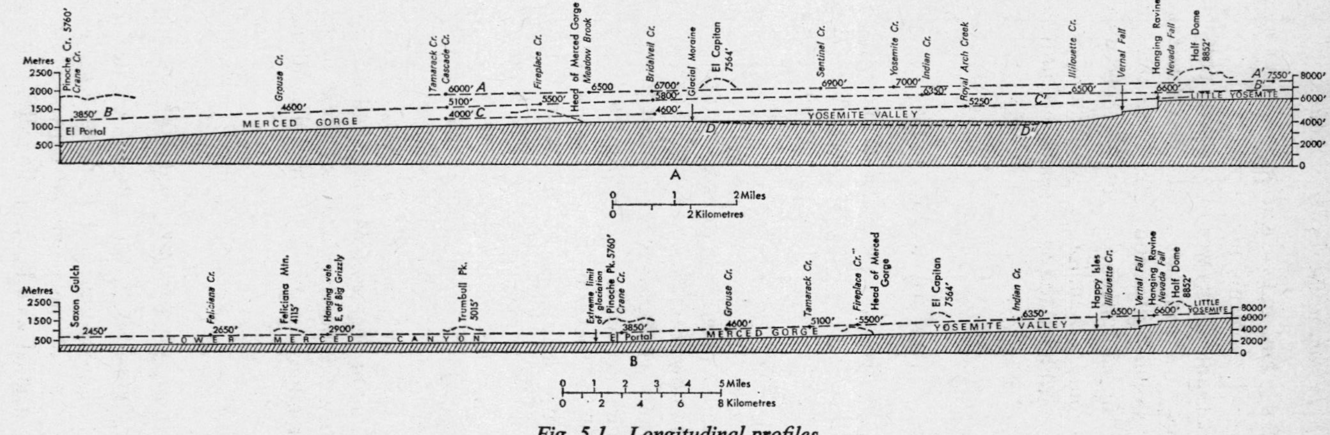

Fig. 5.1 Longitudinal profiles

A. Yosemite Valley, showing three former profiles of the Merced determined by three sets of hanging valleys. *A–A'* is the profile of the earliest or broad-valley stage; *B–B'* is that of the second or mountain-valley stage; *C–C'* is that of the third or canyon stage, which immediately preceded the first ice invasion. The depth of the Yosemite then was 2400 ft, measured from the brow of El Capitan. *D–D'* is the profile of the rock floor of ancient Lake Yosemite, which occupied a basin scooped out by the glaciers. The space between *C–C'* and *D–D'* affords a measure of the depth of glacial and interglacial excavation. This depth ranges from 500 ft at the lower end of the valley to 1500 ft near the upper end. The space between *D–D'* and the present valley profile indicates the depth of the sand and gravel that now fill the basin of ancient Lake Yosemite.

B. Yosemite Valley and part of lower Merced Canyon. The broken line indicates the old profile of the mountain-valley stage, which antedated the last great uplift of the Sierra Nevada and the cutting of the inner gorge. It is determined by the profiles of a number of hanging side valleys, some of which lie far below El Portal and the extreme limit reached by the glaciers of the ice age.

ancient Lake Yosemite) represents excavating done since the beginning of the ice age. True, this is not a measure of glacial cutting alone: it represents glacial cutting and stream cutting combined, for the glaciers were active only at intervals during the ice age, and during the prolonged interglacial stages the Merced River performed its characteristic work. However, it is plain from the very character of the profile *D–D′* and from its extension by the giant stairway to the Little Yosemite and the upper Merced Canyon that glacial cutting was vastly preponderant, for a stairlike canyon profile of this type, characterized by alternating treads and risers – the treads approximately level or bearing shallow lake basins, and the risers cut in the form of precipitous cliffs – is a characteristic product of glacial erosion, not of stream erosion. The tendency of streams is to produce fairly smooth, unbroken valley profiles – to eliminate steps as well as basins. The Merced has done nothing since the ice age to accentuate the stairlike profile of its pathway; on the contrary, it has done what it could do to demolish the steps and fill the basins.

Figure 5.1A shows strikingly the inequalities in the depth of glacial excavation. Evidently the ice accomplished much more work in some places than in others. At the lower end of the valley the glacial deepening measures only about 500 ft, but up the valley it increases gradually, reaching a maximum of about 1500 ft near the head of the valley. Thence it diminishes rapidly to 850 ft at the top of the Vernal Fall and to a minimum of about 250 ft near the top of the Nevada Fall, to increase again gradually to 700 ft in the Little Yosemite. One can not but marvel at these marked inequalities in the deepening effected by the ice – indeed, it is hard to say which seems more astonishing, the maximum or the minimum. However, the true significance of these inequalities becomes apparent only when they are considered together with the variations in the lateral cutting.

WIDENING EFFECTED BY THE YOSEMITE GLACIER

The lateral cutting effected by the Yosemite Glacier is strikingly revealed in the cross profiles in Figs. 5.2–5.8. Nothing, indeed, serves better to give a true conception of the thoroughgoing transformation which the Yosemite Valley has suffered by glaciation than this series of diagrams. Each shows exactly to scale – that is, without vertical exaggeration – the cross profile of a certain part of the valley as it is to-day, plotted from the contours of the topographic map, and the

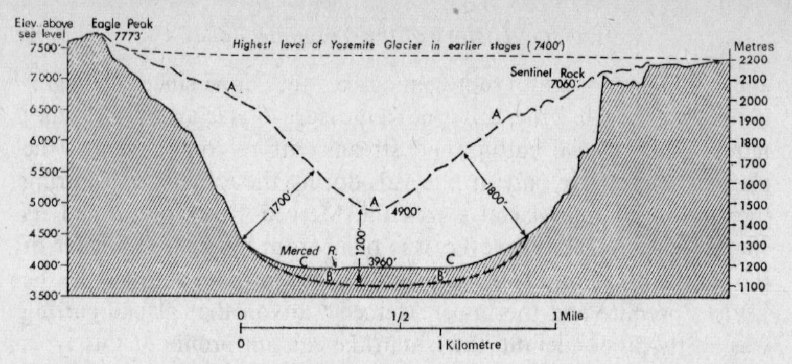

Fig. 5.2 Cross profile from Eagle Peak to Sentinel Rock. Depth of glacial excavation, 1200 ft. Quantity of rock removed by the ice estimated at 1,400,000 yd³. A–A, Preglacial profile; B–B, approximate bottom curve of the glacial U *trough; C–C, present profile*

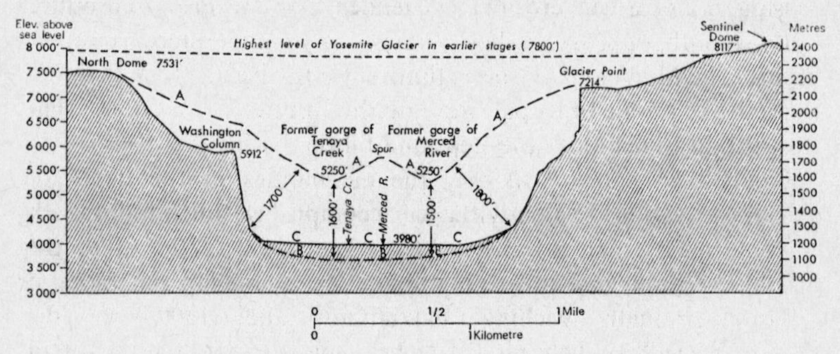

Fig. 5.3 Cross profile from North Dome to Glacier Point, showing maximum depth of glacial excavation, 1500–1600 ft. Quantity of rock removed by the ice estimated at 1,700,000 yd³. A–A, Preglacial profile; B–B, approximate bottom curve of the glacial U *trough; C–C, present profile*

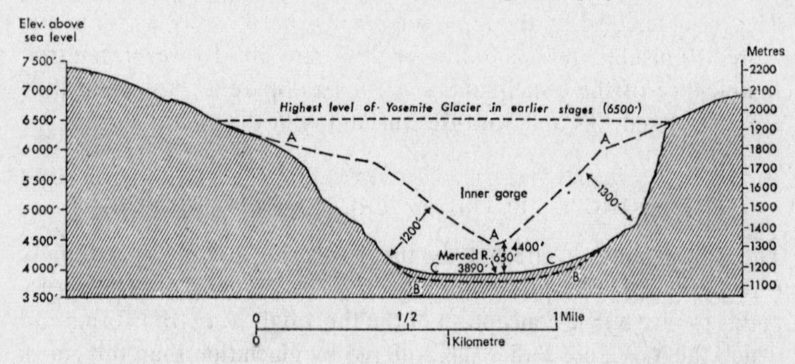

Fig. 5.4 Cross profile below Bridalveil Meadow. Depth of glacial excavation, 650 ft. Quantity of rock removed by the ice estimated at 800,000 yd³. A–A, Preglacial profile; B–B, approximate bottom curve of the glacial U *trough; C–C, present profile*

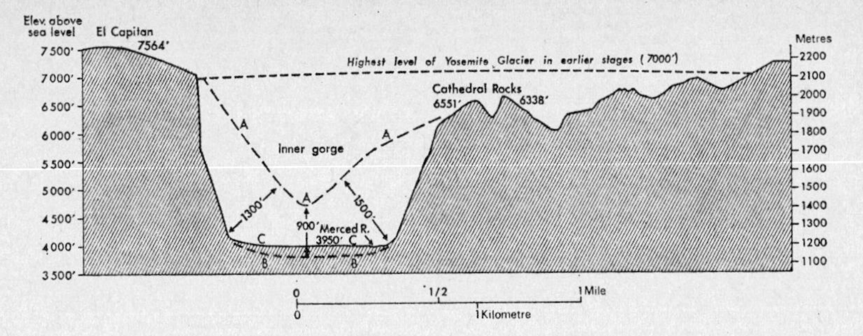

Fig. 5.5 Cross profile from El Capitan to the Cathedral Rocks. Depth of glacial excavation, 900 ft. Quantity of rock removed by the ice estimated at 700,000 yd³. A–A, Preglacial profile; B–B, approximate bottom curve of the glacial U *trough; C–C, present profile*

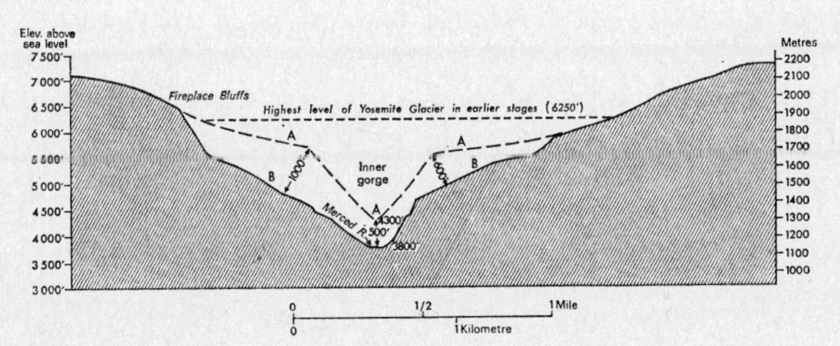

Fig. 5.6 Cross profile at head of Merced Gorge, showing minimum depth of glacial excavation, 500 ft. Quantity of rock removed by the ice estimated at 500,000 yd³. A–A, Preglacial profile; B–B, present profile

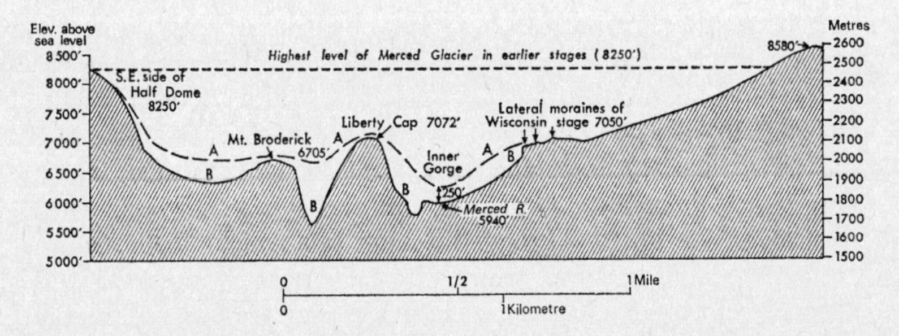

Fig. 5.7 Present cross profile (B–B) at mouth of Little Yosemite Valley compared with the preglacial cross profile (A–A). The changes produced here by glacial action were relatively slight and involved the removal of less than 300,000 yd³ of rock

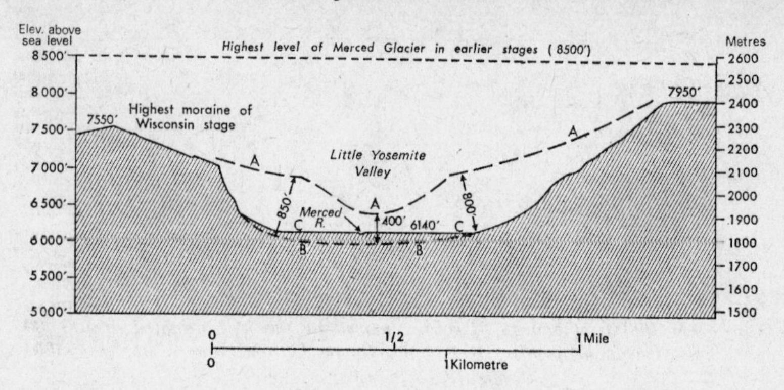

Fig. 5.8 Present cross profile at the middle of the Little Yosemite Valley compared with the preglacial profile (A–A). At least 450,000 yd^3 *of rock was removed from this section by the ice. B–B, Approximate bottom curve of the glacial* U *trough; C–C, profile of the present valley floor on the sediment that fills the lake basin*

corresponding cross profile of preglacial time. The reconstruction of the preglacial cross profile has not been simply a matter of imagination, for the depth of the preglacial Yosemite Canyon has been determined (see profile *C–C′*, Fig. 5.1A), and the character of its sides was governed by a number of elements such as the heights and gradients of hanging valleys and hanging gulches and the forms of truncated spurs and projecting rock monuments. General guidance, further, is afforded by the well-known laws of canyon cutting by streams and of the progressive conversion of sheer canyon walls into slopes of moderate declivity by the processes of weathering and erosion. In some places intimate knowledge of the local structure of the granite has permitted the introduction of sculptural details such as benches, facets, pinnacles and knobs.

From all these cross profiles it is manifest that lateral cutting has been a more important element in the transformation of the Yosemite chasm than downward cutting. At every point the widening accomplished exceeds the deepening. It is, in fact, mainly through lateral cutting that the narrow V canyon of preglacial time has been transformed into the broad U trough of to-day. It is evident, further, from the very breadth of the U profile that the glacial processes far outstripped the fluvial processes, for whatever trenching the river did in interglacial epochs must have tended to produce an inner gorge, yet of such a gorge there is no vestige whatever. Surely it is not to be supposed that the river at any time possessed volume enough to spread over the whole width of the valley and that it performed a significant share of the lateral cutting. If it had had such volume and power it would have produced many broad-bottomed yosemites in the relatively unresistant sedimentary rocks below El Portal long before it evolved a single valley of that type in the resistant granitic rocks of the Yosemite region.

Comparison of the different cross profiles with one another reveals also the fact that the widening accomplished was no more uniform than the deepening and that in general the glacial processes worked very unequally. They accomplished large results in each of the two chambers of the Yosemite Valley, but considerably less in the portal between those chambers and astonishingly little in the Merced Gorge below the valley and in the gorge below the mouth of the Little Yosemite.

A fair measure of the lateral cutting effected in the main Yosemite chamber is indicated in the cross profile from Eagle Peak to Sentinel Rock shown in Fig. 5.2. Measured at right angles to the sides of the preglacial canyon it amounts to fully 1800 ft, whereas the deepening amounts to approximately 1200 ft. The space between the preglacial and postglacial profiles as shown in the diagram measures about 1,400,000 yd^2; hence the ice excavated from a section of the valley 1 yd long in the direction of its axis about 1,400,000 yd^3 of rock. This measurement, it is true, is made at a point where a jagged spur projected from the face of Sentinel Rock, yet the amount stated may be considered fairly representative of most of the main Yosemite chamber, for there were several other spurs, all of which were planed away by the glacier. The amount is of course merely an approximation, but it will serve for comparison with similar amounts indicated by other cross profiles.

The maximum of glacial excavation is shown in the cross profile from North Dome to Glacier Point and Sentinel Dome (Fig. 5.3). The ice there accomplished truly prodigious results: at no other place did it effect a more complete transformation. The conditions were, however, peculiar to the head of the valley, for the two preglacial gorges of Tenaya Creek and the Merced River there came together, and the two main branches of the Yosemite Glacier were confluent. The hump in the middle of the profile represents in section the tapering spur that sloped formerly from the west base of Half Dome to the junction of the gorges. A more massive spur doubtless projected northward from Glacier Point, and a smaller spur projected from North Dome in the direction of the Washington Column. The quantity of rock excavated by the ice from a section of the valley 1 yd long in the direction of its axis, computed from this diagram, approximates 1,700,000 yd^3.

Representative of the lower Yosemite chamber is the cross profile shown in Fig. 5.4. Here too, unquestionably, entire spurs were removed by the ice. The lateral cutting, measured as before at right angles to the preglacial canyon sides, amounts to nearly 1400 ft, and the total quantity of rock excavated over the distance of 1 yd is about 800,000 yd^3. The cross profile of the portal between El Capitan and the Cathedral Rocks (Fig. 5.5), on the other hand, shows only 700,000 yd^3 of excavation. But the most marked contrast is offered by the profile in Fig. 5.6, which is taken across the Merced Gorge, just below the Yosemite Valley. The work done there by the ice can scarcely be called a transformation. There was only moderate enlargement of the valley section and no real change in general form, the inner gorge, though widened somewhat, remaining strongly in evidence. The quantity of rock excavated per yard here is less than 500,000 yd^3 – that is, little more than one-half the quantity indicated for the lower chamber and little more than one-third the quantity indicated for the upper chamber.

Another cross profile of special interest is that in Fig. 5.7, drawn across the mouth of the Little Yosemite and over Mount Broderick and Liberty Cap. It shows an abrupt diminution in the glacial excavation, the quantity removed from a section 1 yd long being only about 300,000 yd^3. Of course this quantity is not comparable on even terms with the quantities shown in the cross profiles previously considered, for it represents work done only by the Merced Glacier, not by the entire Yosemite Glacier. But even for the Merced Glacier the quantity stated is very small. The profile in Fig. 5.8, which is taken across the middle of the Little Yosemite, shows an amount of material excavated more than half as large again.

At the head of the Little Yosemite, finally, glacial excavation dwindles to a minimum. There, opposite Bunnell Point, the walls contract and the upper gorge of the Merced begins – a gorge so narrow that, were it not for the smoothness of its walls and the presence of polish, striae, and grooves, it might readily seem a product of stream erosion purely. To obtain a true conception of that gorge in its relations to the upper Merced Canyon as a whole, one should view it from a lofty summit such as Half Dome or Clouds

Rest. It is then seen to be a mere inner trench winding across a large billowy mass of bare granite that obstructs the canyon for a distance of 2 miles.

How are these extreme and in places abrupt variations in the depth and breadth of glacial excavation to be explained? Why have the glaciers been able to accomplish so much in some parts of the Yosemite region and so little in others?

A comparative study of the principal glacial troughs of the Sierra Nevada shows that they are by no means proportionate in size to the glaciers that occupied them. Some of the greatest were the pathways of only moderately large glaciers, and some of the lesser were the pathways of very large glaciers. Thus the Yosemite itself, the most capacious of all the valleys of its type, was the pathway of a glacier only 37 miles long and slightly over 3000 ft thick at the time of maximum glaciation; whereas the Hetch Hetchy Valley, though only half as long and half as wide as the Yosemite and about 1000 ft shallower, was traversed by a glacier 60 miles long and 4000 ft thick. [To explain this paradox demands that we consider] another factor – a factor that in large measure controlled the glacial processes and determined how much excavational work they might accomplish at any place. That factor is the structure of the rock.

In order that the influence of rock structure on the excavating efficiency of glaciers may be understood it is desirable first to gain a clear idea of the manner in which glaciers do their work.

HOW DO GLACIERS EXCAVATE?

It is commonly supposed that glaciers erode their beds mainly by grinding and scouring. It is true that with the rock fragments which they hold in their bottom layers glaciers perform considerable abrasive work – witness the polished, striated and even deeply grooved floors and walls of glaciated canyons. Nevertheless, the efficacy of this abrasive action is not inherently great. Only in soft, friable rocks does it accomplish really large results. In hard, tough rocks, such as granite, it achieves but little – not enough, in any event, to account for the profound remodeling of entire valleys and canyons. On such rocks, as a rule, the presence of glacier polish is indicative of moderate changes slowly produced.

The process whereby glaciers excavate to best effect in hard rocks is by plucking, or 'quarrying' entire blocks and slabs. Because of their very weight – some 30 tons to the square foot for every thousand feet of thickness – and the fact that they are shod with coarse rock waste frozen in their basal layers, glaciers have a strong frictional hold on their beds; and so, as they move forward, though at a rate of only an inch or two a day, they dislodge and drag forth entire blocks and slabs. The peculiar property that ice has of freezing tightly to objects with which it is in contact is probably a potent factor in the process.

The blocks and slabs thus dislodged are, however, rarely broken off from sound, unfractured rock. The glaciers take advantage, rather, of the fractures already existing in the rock – the joints by which it is divided into natural blocks and slabs. This is true especially of hard, tough rocks such as granite, for even a glacier 3000 ft thick does not exert pressure enough to disrupt a floor of sound, massive rock of that type. It is clear, then, that the joint structure plays a very important part in glacial quarrying. Without it, in fact, the process is scarcely operative in hard rocks.

Several agencies, furthermore, tend to facilitate the quarrying process, by loosening up the blocks and slabs. Acid carried by water that percolates through the joint fractures dissolves the weaker minerals and lessens cohesion; and water freezing in the fractures pushes the blocks and slabs apart with its momentary but strong expansive force. It has even been contended, by those who would attribute only slight erosive power to glaciers, that the quarrying is limited practically to the removal of blocks and slabs previously loosened in preglacial or interglacial time, but there is ample evidence to the contrary. In many glacier channels the quarrying can be seen to have progressed far below the zone of weathering, its depth varying primarily with the thickness and power of the glacier concerned.

The manner in which the glacial quarrying process operates is illustrated by Fig. 5.9. Any joint block in the bed of a glacier, such as that marked A, which is for any reason unsupported or weakly supported on its downstream side, is particularly susceptible of being dislodged, for the force of the glacier is exerted upon it at a small angle forward from the vertical, as indicated by the arrow. The block A and its side companions having been removed, the block B and those flanking it will next be unsupported and ready for removal, and so the process will continue farther and farther up the valley. Its rate of progress will depend upon the power of the glacier, the size and weight of the blocks, and the looseness of the joints.

Impressive evidence of the quarrying action of glaciers is to be found also in the presence of angular, sharp-edged blocks in the moraines. Such blocks abound especially along the upper courses of

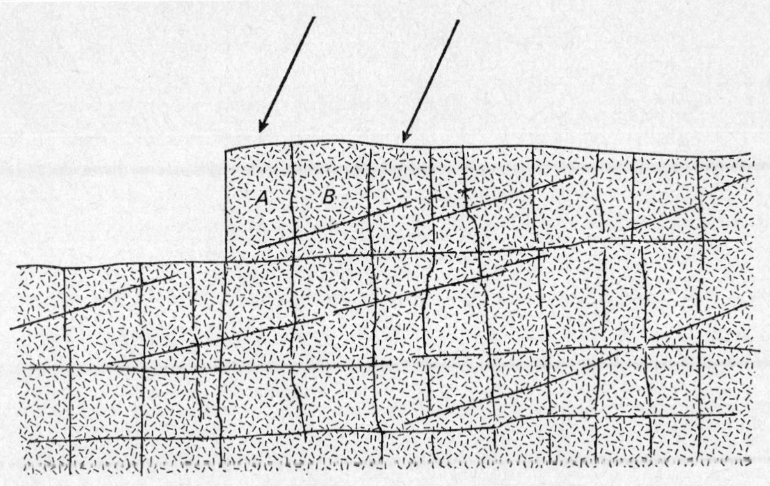

Fig. 5.9 Diagram illustrating the quarrying of joint blocks by a glacier. The arrows indicate the direction in which the ice exerts its pressure, the movement being from right to left

lateral moraines, where they were dropped before they had been carried far enough to lose their angularity by wear.

SELECTIVE QUARRYING

Because glacial quarrying in hard rocks is so largely dependent upon the presence of joints, its action naturally is strongly influenced by the spacing of the joints. Where these fractures are close together, so that the rock is divided into small, light blocks or slabs, quarrying will proceed with relative ease and rapidity; there even a small, feeble glacier will be able to accomplish much. Where the joints are far apart – 50 ft or more – the blocks between them are too large and too heavy even for a mighty trunk glacier to dislodge. Rock so sparsely jointed is virtually massive, so far as the glacial processes are concerned, and can be reduced only by slow abrasion.

Between these two extremes, of course, there are many intergradations, depending upon the distribution and the arrangement of the fractures. In these intermediate rocks in which the jointing is moderately coarse the excavating force of the glacier is as a rule the decisive factor. In dealing with rocks so jointed a small shallow glacier exerting relatively little pressure on its bed may be almost impotent, whereas a glacier of great depth, exerting correspondingly great pressure, may accomplish signally large results. It is, in fact, precisely in those areas where the joint blocks are fairly large – from 5–25 ft to the side – that a powerful glacier may attain its greatest excavating efficiency, for every block it removes has great cubical content. And this explains why in regions of coarsely jointed rocks there is usually a vast difference between the amount of excavational work done by the great trunk glaciers and that done by the small tributary glaciers.

This brief analysis shows that the quarrying action of glaciers is inherently selective, especially in regions where the rocks are hard and tough. There its effectiveness is dependent in large measure upon the character of the rock structure – more specifically upon the spacing of the joints.

Bearing this in mind let us now briefly examine the jointing of the rocks of the Yosemite region and see in what manner it has affected the action of the glaciers. No one who gives the rock walls of the Yosemite, the Little Yosemite and Tenaya Canyon more than a superficial glance can fail to note the marked variations that occur in their joint structure. Not only does the arrangement of the joints differ from place to place, but the spacing varies widely. What is more, these variations are sometimes remarkably abrupt, so that structural extremes are brought into immediate juxtaposition. In few other regions where granitic rocks occur is there so great structural diversity or are sharp contrasts in structure so prevalent as in the Yosemite region.

Now the course of the Merced in its larger aspects is by no means related to or controlled by these structural vagaries in the granitic rocks; it is essentially a 'superimposed' stream, and so are most of the other rivers on the western slope of the range. Therein lies the key to the secret of the origin of the Yosemite Valley and of all the other yosemites in the Sierra Nevada. Each of these capacious U-shaped valleys has been developed in an area of prevailingly fractured rocks in which the agents of erosion worked with comparative facility and in which the glaciers, when they came upon the scene, quarried with extraordinarily great effect. The narrow portals and gorges above and below the yosemites, on the other hand, are cut in bodies of prevailingly massive rock which the glaciers could not quarry and could reduce only by slow grinding.

CONFIGURATION OF THE YOSEMITE VALLEY EXPLAINED

The central part of the main Yosemite chamber has been excavated from a body of coarsely jointed granodiorite – the Sentinel granodiorite. This rock extends directly across the valley in a belt about 2 miles wide. The Three Brothers and Taft Point mark its western margin; the Royal Arches and Glacier Point its eastern margin. It is divided mainly by vertical and horizontal joints and hence has in many places a distinctly columnar or prismatic structure. This is evident especially in the columnar crags and pinnacles on the wall east of Union Point. In other places, especially in Sentinel Rock, it has a smoothly sheeted structure, the partings being nearly vertical. Almost throughout, therefore, this rock material is divided into blocks and sheets of large size – of a size, in fact, which the Yosemite Glacier during its higher stages could quarry with great efficiency.

From the vicinity of Taft Point west as far as the Cathedral Rocks diorite and gabbro predominate. These of all the rocks of the Yosemite region are the most thoroughly fractured; hence they must have been readily quarried by the glacier, even at times when it had only moderate volume. It is not surprising, therefore, to find that in the area of these rocks the south wall is embayed. Just how far into the valley these well-jointed rocks extended originally can only be surmised, but there is reason to believe that they occupied considerable space, for the valley here attains its greatest width, in spite of the fact that its north wall is composed of prevailingly massive granite.

The narrow portal between El Capitan and the Cathedral Rocks is, as might be expected, framed by promontories of exceptionally massive rock that could not be quarried by the glacier. The great prow of El Capitan consists wholly of this highly siliceous, massive granite – El Capitan granite it has been appropriately named. The Cathedral Rocks consist only in part of granite of this type and are traversed by numerous sheets and dikes of other igneous rocks, but they are nevertheless for the most part unfractured, the different rock materials in them being intimately welded together. But for this fact the whole promontory surmounted by the Cathedral Rocks and bearing the hanging gulch of Bridalveil Creek would probably have been quarried away by the glacier flush with the south wall of the valley.

The lower Yosemite chamber doubtless owes its great width to the ease with which the glacier quarried in the large bodies of well-jointed gabbro and diorite which extend throughout most of its length. Considerable masses of these dark-hued basic rocks still cling to the north side, west of the Ribbon Fall. Their unstable masonry, crisscrossed by numerous joints, has not remained standing in the form of a sheer wall but has broken down completely, producing the immense talus known as the Rock Slides, over which the Big Oak Flat Road is built.

On the south side glacial quarrying has been less effective, the bulk of the rock being El Capitan granite. As a consequence sheer walls and massive buttresses remain, but there are several recesses which show that the quarrying has been facilitated and guided locally by zones of intense jointing and shattering. The recess dominated by the Leaning Tower is of this kind. It is entirely probable, further, that the abrupt increase in the width of the valley below the portal and the persistent southward trend of the wall extending from the Bridalveil Fall to the Leaning Tower are due in large part to the influence exerted by the same zone of shattering on the glacial quarrying.

Significantly the lower Yosemite chamber contracts abruptly west of the body of fractured gabbro and diorite, and the great barrier which incloses the spoon-shaped lower end of the valley consists of massive El Capitan granite. The benches that flank the Merced Gorge as far west as the Gateway are composed of the same obdurate material, a fact which accounts for the narrowness of the gorge. Close examination shows, however, that this granite is not wholly massive but traversed at long intervals by vertical and horizontal master joints. It is therefore really divided into blocks, but these were much larger than the glacier could dislodge. The ice merely rode over them, grinding their surfaces, as is strikingly revealed by the smoothly curving shieldlike hump of Turtleback Dome.

The lower end of the Yosemite Valley, though scenically unattractive, is of peculiar scientific interest. Few other localities in the Yosemite region afford more striking evidence of the dependence of glacial quarrying upon the presence of favorable structures in the rock and of the comparative inefficiency of glacial abrasion in massive rock. Though the abrupt contraction of the valley at its lower end might at first sight seem to indicate the place where the glacier usually terminated and beyond which it only rarely advanced, it marks in fact but the western limit of the quarriable rocks in the

Yosemite Valley and the beginning of the unquarriable rocks along the Merced Gorge.

The lower end of the valley is of interest, further, because there a considerable share of the glacier's mass had to move upward in order to get out. The central portion of the glacier, of course, passed through the Merced Gorge without moving upward, but the flanking masses rode up the rock slope at the end of the valley and surmounted the uneven benches that flank the gorge on both sides. Indeed, the deeper the glacier excavated the Yosemite Valley, the higher these ice masses had to climb in order to make their exit. Toward the end of the earlier stages of glaciation they had to climb 1000 ft. This almost incredible ascent of the ice is attested beyond possible doubt by the striae and associated glacial markings on the rock slope. Of greatest value as indicators of the direction of ice movement are the so-called 'chatter marks', of which a few can be distinguished here and there. These are fine curving tension cracks in the rock produced by heavy boulders that were dragged by the glacier, and they are invariably bowed upstream.

Three other facts of prime importance remain to be explained – the great depth to which the head of the valley has been excavated below the level of the preglacial gorge, a depth not less than 1500 ft, it would appear from the longitudinal profiles in Fig. 5.1A; the steady decrease in the depth of glacial cutting from the head of the valley down to the lower end, where it measures only about 500 ft; and the scooping out of the basin of ancient Lake Yosemite in the rock floor of the valley. These matters, however, are all bound up with the question, how are the stairlike steps with basined treads characteristic of profoundly glaciated canyons produced?

ORIGIN OF GLACIAL STAIRWAYS

The giant stairway from whose main steps the Vernal and Nevada Falls descend, impressive though it may be, taken by itself, is after all only the beginning of a much longer stairway that extends throughout the upper Merced Canyon from the Yosemite Valley to the base of Mount Lyell – a stairway 21 miles in length and making a total ascent of 7600 ft. Some of the steps in that greater stairway are ill formed, none are as clean-cut as those at the Vernal and Nevada Falls, and nearly every one of them has a shallow basin hollowed out in its tread, yet the stairlike character of the canyon profile as a whole is unmistakable. Moreover, the floors of the main Yosemite and Little Yosemite are seen to constitute treads in that greater stairway. They do not differ materially from the other treads save in their greater length and in the fact that the basins in them are completely filled with stream-borne sediment, whereas on the upper treads the basins are filled only in part. The rock floor of the Yosemite Valley really comprises two treads differing but slightly in altitude – a short one in the lower chamber and a long one in the upper chamber. The rock sill on which the moraine dam at the El Capitan Bridge rests forms the edge of the upper tread.

Such stairwise ascent by successive steps has long been recognized as a characteristic feature of strongly glaciated canyons, but the precise nature of the process whereby such canyon steps are produced is still a moot question. Several different hypotheses have been offered in explanation, such as that of Willard D. Johnson (1904), [but none has proved entirely satisfactory].

Observations carried over a considerable part of the Sierra Nevada show that most steps have risers and sills composed of very sparingly jointed or wholly massive rock. Even in the areas of sedimentary and volcanic rocks the canyon steps have as a rule sills of more than ordinarily resistant materials. Particularly instructive in this regard is the flight of steps in Bloody Canyon, on the east flank of the Sierra Nevada, which is composed of a variety of rocks, sedimentary, volcanic, and granitic. In nearly every step the slightly raised edge or sill, from whatever material it may be hewn, is associated with a constriction in the canyon section due to the resistance offered to glaciation by the same obdurate rock in the sides of the canyon. Such a constriction is to be seen at the Vernal Fall; and another is to be seen at the top of the giant stairway, where the mouth of the Little Yosemite is partly blocked by the obdurate masses of Liberty Cap and Mount Broderick. The treads, on the other hand, are almost invariably broad. They are the broadest parts of the canyons and, with their embayed sides and concave floors, constitute roughly spoon-shaped basins situated at successive levels one above another.

It must be clear to anyone who considers these facts that rock structure, or, more broadly, rock resistance, plays an important part in the development of canyon steps by glaciation; that, indeed, it determines in large measure at what points in a given canyon the individual sills and treads shall develop. That being so, it follows that no hypothesis that aims to explain the production of glacial stairways can be considered satisfactory that fails to take into account this influence of rock resistance. Accordingly, a new explanation suggests itself – an explanation that is in harmony with the principle of selective quarrying already laid down.

Briefly stated, this new explanation is as follows: In a canyon or valley cut in granitic rocks of widely varying structure such as prevail in the Sierra Nevada, a glacier is bound to excavate with locally

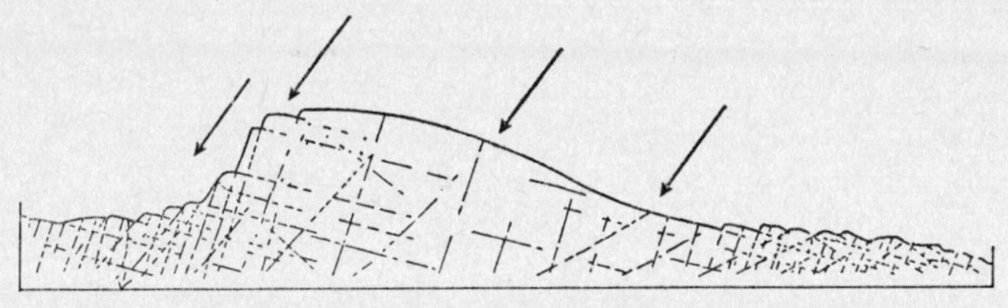

Fig. 5.10 Longitudinal section of a typical roche moutonnée fashioned by a glacier from an obdurate mass of sparsely jointed granite. The glacier moved from right to left and exerted its force in the direction indicated approximately by the arrows – that is, at a high angle against the back and crown of the hump but at a slight angle away from the downstream face. It consequently subjected the back and crown to vigorous abrasion, leaving them smoothed and gently curved, and it subjected the downstream face to quarrying mainly, leaving it hackled and abrupt. If glaciation had continued until all of the jointed, quarriable rock had been removed from the downstream side, there would have resulted an asymmetric dome, smoothed on all sides but steeper on the downstream side than on the upstream side. An example of such a completely smoothed roche moutonnée of massive granite is the small nameless dome that stands in the Little Yosemite about half a mile northeast of Liberty Cap

varying efficiency; where the rock is massive or only sparsely divided by fractures, the glacier, being unable to disrupt the rock, can reduce it only by abrasion – a slow and relatively feeble process; on the other hand, where the rock is plentifully divided by natural partings, the glacier will quarry out entire blocks and excavate at a fairly rapid rate. From the first, therefore, it will tend to work irregularly, producing hollows in the areas of jointed rock and leaving obstructing humps in the areas of massive, unquarriable rock. The humps, however, will tend to assume strongly asymmetric forms, gently sloping and smooth on the upstream side, abrupt and more or less hackly on the downstream side, for, as will be seen in the diagram in Fig. 5.10, the force of the ice is directed at an angle against the upstream side and so subjects it to intense abrasion; and the force is directed at a small angle away from the downstream side and so exerts there a pull favorable for quarrying.

The smaller knobs of this asymmetric type have come to be known by the quaint name *roches moutonnées*, which was given to them by the Swiss mountaineers because, when viewed from up the valley, their rounded forms suggest the backs of grazing sheep. That they owe their peculiar modeling to abrasion on the upstream side and quarrying on the downstream side is quite generally recognized, but that the larger obstructions which occupy the entire breadth of the canyon floor and form the sills and edges of the steps are shaped in essentially the same way appears not to be generally understood. In the upper Yosemite region and the adjoining parts of the High Sierra, however, moutonnée forms of all sizes abound, ranging from mere hillocks and rock waves a few feet high to canyon steps a thousand feet high, and there the inherent kinship of all these features is readily manifest to one who takes the trouble to compare them with one another.

A typical example that is intermediate between a mere roche moutonnée and an entire canyon step and is readily accessible for inspection is the abrupt rise in the canyon floor halfway between the Vernal Fall and the Nevada Fall, which was the site of the historic hostelry known as La Casa Nevada. It has been referred to as the second step of the giant stairway, but it extends really only part of the way across the canyon. The trail leads steeply up to the top of the sill through a notch in the downstream face, which is determined by strong vertical master fractures, and then it leads down again along the gentle back slope of sparsely fractured granite.

The manner in which the basined treads of a glacial stairway are evolved will be most readily understood if the treads are viewed in reference to the slope of the preglacial canyon floor. If the diagram in Fig. 5.11 is so tilted as to make the preglacial canyon floor appear horizontal, the treads will assume the aspect of basins that are strongly asymmetric, being deepest at their upper ends. They are so shaped manifestly because the ice erodes with greatest vigor at their upper ends; it is thicker there than at their lower ends, descends into them with plunging motion, and at the foot of its cascade is compelled to make an abrupt turn, as is indicated by the arrow – circumstances all of which cause the ice to exert particularly great pressure on its bed. Downstream, of course, the pressure diminishes progressively, reaching a minimum at the edge of the step.

Treads are shaped both by quarrying and by abrasion, but it is a fair presumption that the quarrying process is dominant wherever the rock is jointed and the glacier has sufficient power to dislodge the blocks. In Fig. 5.11 several stages in the evolution of a tread are shown in order to bring out the fact that, as the quarrying proceeds, always in the headward direction, numerous minor cross cliffs controlled by joints are likely to be developed. But these are only temporary hackles in the canyon floor that migrate headward and are eliminated in the course of time. The main canyon steps, on the other hand, are seen to be fairly stable features that are cut back only very slowly, their sills and risers being composed of the more massive and

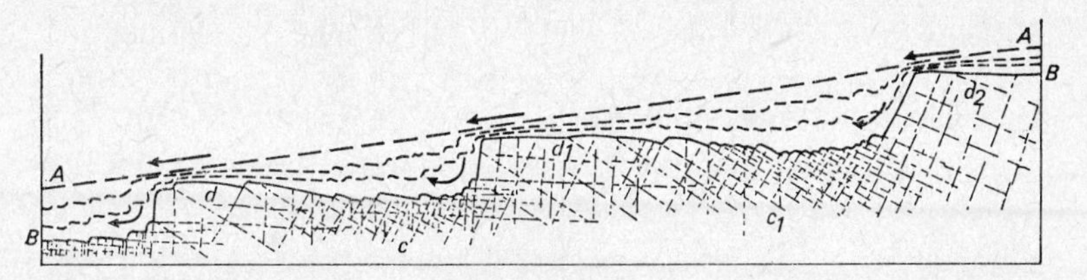

Fig. 5.11 Longitudinal section of a canyon illustrating the mode of development of a glacial stairway by selective quarrying. AA represents the profile of the preglacial canyon floor; BB that of the glacial stairway. Bodies of closely jointed rock, such as c and c_1, are readily quarried out by the glacier, but bodies of sparsely jointed, unquarriable rock, such as d and d_1, being reducible only by abrasion, remain standing as obstructions with flattened and smoothed tops and steep, more or less hackled fronts. The broken lines indicate successive stages in the development of the steps and treads. The arrows indicate the direction of ice movement

therefore more obdurate rock. Likewise they are worn down very slowly – mainly by abrasion – and the sill of each step therefore controls the general level of the tread back of it. No other hypothesis advanced hitherto has explained what determines the level at which a tread shall be developed.

It is evident, further, that the process is a self-intensifying one, which constantly tends further to accentuate the stairlike profile of the canyon. For the greater the depth of excavation at the foot of a given step becomes, the more powerful will be the glacial action there. The limits to which the process can go in any locality are determined, of course, by the force of the glacier, the length of time it is active, and the resistance of the rocks. Finally, there is no inherent tendency in the process to produce level or nearly level treads; it works regardless of gradients. It may produce treads that are approximately level, or that have a gentle slope, either forward or backward, or that have one or several basins scooped out in them, all depending upon the structure of the rocks involved and the degree to which that structure controls the glacial action.

SUMMARY OF CHANGES PRODUCED BY THE GLACIERS

The changes that were brought about in the configuration of the Yosemite region by the repeated ice invasions of the glacial epoch may be summed up as follows:

In the Yosemite Valley itself, where the rocks were prevailingly well jointed, glacial quarrying was particularly effective and accomplished conspicuously large results. Both downward and sideward the Yosemite Glacier quarried, trimming off projecting spurs, cutting back the craggy slopes of the preglacial river canyon to sheer, smooth cliffs, and transforming the brawling cascades descending from the hanging side valleys into leaping falls of astounding height. Throughout the length of the valley the inner gorge of the Merced was wiped out of existence and in its stead there was produced a broadly concave, basin-shaped rock floor. Even the features of the mountain valley of the Pliocene epoch were largely destroyed, and thus from a tortuous V canyon the Yosemite was enlarged to a spacious, moderately sinuous U trough with approximately parallel, spurless sides and with a long lake basin in its bottom (Fig. 5.12). Only between El Capitan and the Cathedral Rocks, composed of massive rock which the glacier was unable to quarry away, did it leave a marked constriction. In the areas of sparsely jointed granite immediately above and immediately below the Yosemite Valley the ice was able to effect but moderate changes in the form of the canyon. There in consequence the inner gorge of the Merced remains preserved and still presents the characteristics of a trench worn by the river in the bottom of an old mountain valley.

In the Yosemite Valley the depth of glacial excavation decreases from a maximum of about 1500 ft at the head to a minimum of about 500 ft at the lower end. The reason for this decrease in glacial deepening down the valley is found in the fact that during all phases of glaciation the ice was thicker and therefore had greater excavating power at the head of the valley than at the lower end, and during the maximum phases it plunged into the head of the valley in the form of a mighty cataract.

Above the Yosemite Valley the Merced Glacier sculptured the steps of the giant stairway from local bodies of extremely massive

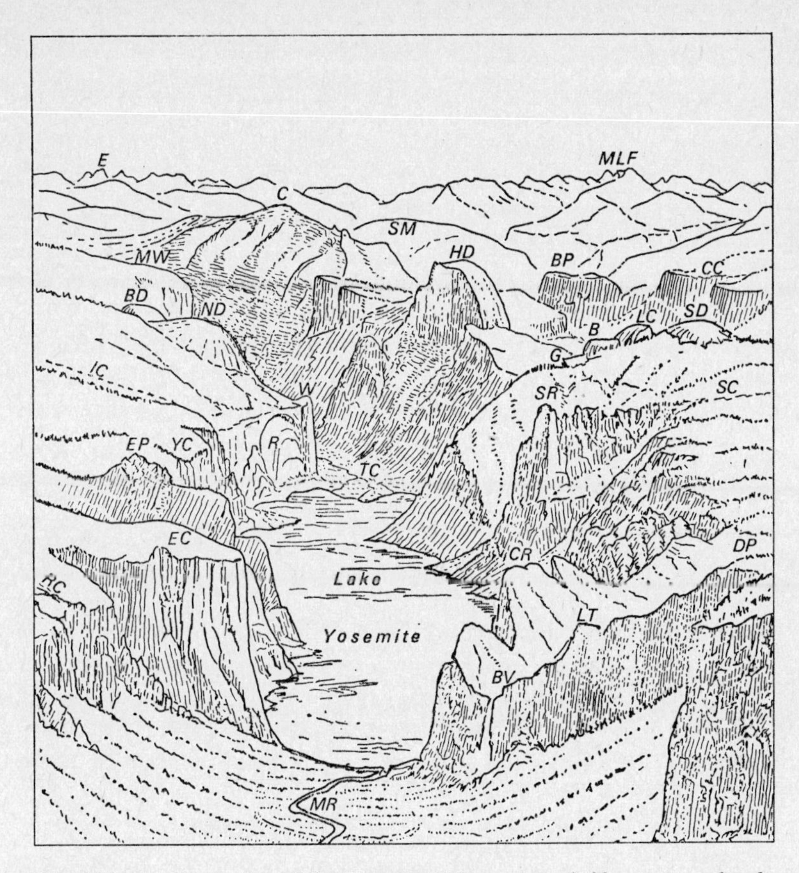

Fig. 5.12 Bird's-eye view of the Yosemite Valley as it probably was immediately after the ice age. The valley had been broadened and deepened to essentially its present proportions. The deepening accomplished by the ice ranged from 600 ft at the lower end to 1500 ft at the upper end. A third set of hanging valleys had been added, and the Bridalveil Fall was produced. A lake 5½ miles long occupied a basin gouged into the rock floor of the valley and dammed in addition by a glacial moraine. The vegetation consisted in the main of types now prevailing. The drawing is based primarily on a systematic survey of the moraines and the other glacial features of the valley

RC Ribbon Creek
EC El Capitan
EP Eagle Peak
YC Yosemite Creek
IC Indian Creek
R Royal Arches
W Washington Column
TC Tenaya Creek
ND North Dome
BD Basket Dome
MW Mount Watkins

E Echo Peak
C Clouds Rest
SM Sunrise Mountain
HD Half Dome
M Mount Maclure
L Mount Lyell
F Mount Florence
BP Bunnell Point
CC Cascade Cliffs
LY Little Yosemite Valley
B Mount Broderick

LC Liberty Cap
SD Sentinel Dome
G Glacier Point
SR Sentinel Rock
SC Sentinel Creek
CR Cathedral Rocks
BV Bridalveil Creek
LT Leaning Tower
DP Dewey Point
MR Merced River

rock. By quarrying headward directly up to the vertical master joints delimiting those bodies and by grinding them down from above, it gave them their marvelously clean-cut, steplike forms.

In the lower half of the Little Yosemite lateral quarrying was favored but downward quarrying was impeded by the structure of the rocks, and as a consequence that part of the valley was given great breadth but relatively little depth. Liberty Cap and Mount Broderick, being composed almost wholly of massive, unquarriable rock, were left standing at its mouth as two roches moutonnées of gigantic size. The upper half of the Little Yosemite has remained narrower than the lower half because it is hemmed in by extensive bodies of massive rock. The head of the Little Yosemite was, like the head of the main Yosemite, the site of a powerful ice cataract and to that circumstance largely owes its great depth and peculiar configuration. In the Little Yosemite as in the main Yosemite the depth of glacial excavation decreases steadily down the valley and for the same reasons.

In Tenaya Canyon, downward quarrying was favored by the presence of a longitudinal belt of fractures, but lateral quarrying was restricted by flanking masses of undivided rock. As a consequence that canyon was cut down almost to the level of the Yosemite Valley and was given even greater depth than that valley, but it was left comparatively narrow and in part shaped like a sharp-keeled boat rather than like a U trough. Only an imperfect stairway was developed in its floor, but at its head a gigantic step was fashioned from a body of massive granite under the cascading action of the Tenaya Glacier.

The Yosemite, the Little Yosemite and Tenaya Canyon, then, owe their present configuration very largely to glacial action. All three had been cut in preglacial time to considerable depth by the streams flowing through them and had acquired the aspect of rugged V canyons, but so thoroughgoing was the remodeling action of the glaciers that now only a few vestiges of their preglacial forms remain. Each chasm had a glacial cataract at its head, yet each differs from the other two in general shape and proportions because in each the structure of the rocks was different and influenced the glacial processes in a different way.

6 Observations on the Drainage and Rates of Denudation in the Hoffellsjökull District

SIGURDUR THORARINSSON

IN the summer of 1936 I joined the Swedish–Icelandic Vatnajökull Expedition, led by H. W:son Ahlmann and Jón Eythórsson. When they had left Hornafjördur on 21 June, Carl Mannerfelt and myself continued the ablation measurements on Hoffellsjökull until the middle of August. Besides these ablation measurements, we took the first observations of Hoffellsjökull's movements, and samples of the water in Hornafjardarfljót to determine its silt content.

Compared with the main work of the Swedish–Icelandic Vatnajökull Expedition, the observations of the glacier movements and the silt content of the rivers were carried out rather unsystematically. Being the first of their kind in these districts, they are nevertheless of some value, as they provide a basis and some suggestions for continued investigations.

Hoffellsjökull (Fig. 6.1) is a typical separate glacial drainage area of the kind usual on the south-east and east sides of Vatnajökull and represented by the 'Big Five' outlet-glaciers: Hoffellsjökull, Fláajökull, Heinabergsjökull, Breiðamerkurjökull and Skeiðarárjökull. These glaciers all flow from the central ice plateau through a flat, spoon-shaped basin, which we have called the intake zone, into more or less pronounced valley glaciers, the lowest parts of which reach the coastal plain, where they fan out slightly.

Fig. 6.2 is a longitudinal profile of the lower part of Hoffellsjökull along the median line of the glacier. The glacier surface and the topography of the background are based on Steinthór Sigurdsson's map from the autumn of 1936, scale 1:50,000, and on my own studies of the ice-dammed lakes.

Hoffellsjökull is drained by two rivers, or river systems, Austurfljót and Vesturfljót. Of these, Austurfljót is by far the biggest, and drains almost the whole of Hoffellsjökull proper (Fig. 6.1). To the two rivers has been given the common name of Hornafjardarfljót. The Austurfljót – the main arm of Hornafjardarfljót – may be regarded as a typical example of the sandur-forming glacier rivers emanating

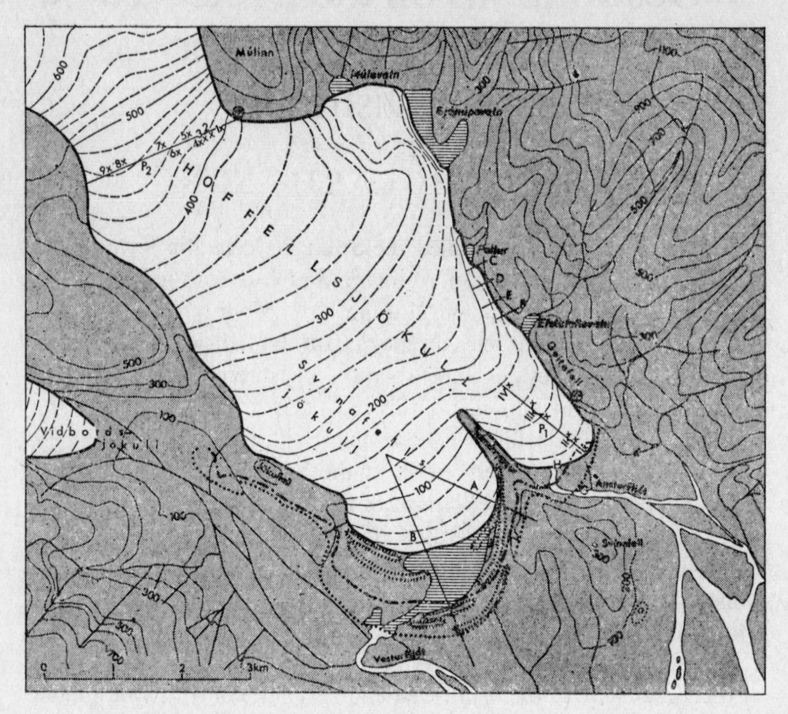

Fig. 6.1 The lower part of Hoffellsjökull. Contours in metres by Steinthór Sigurdsson, 1936. The outer dotted line indicates the ice front about 1890, the inner dot-and-dash line the front in 1903 (according to the topographical map of the Danish General Staff). P_1 and P_2: movement-profiles; crosses mark the points of which the rate of movement was measured, encircled crosses base points. A–F: profiles for measuring the thinning. G–L: places where water samples were taken

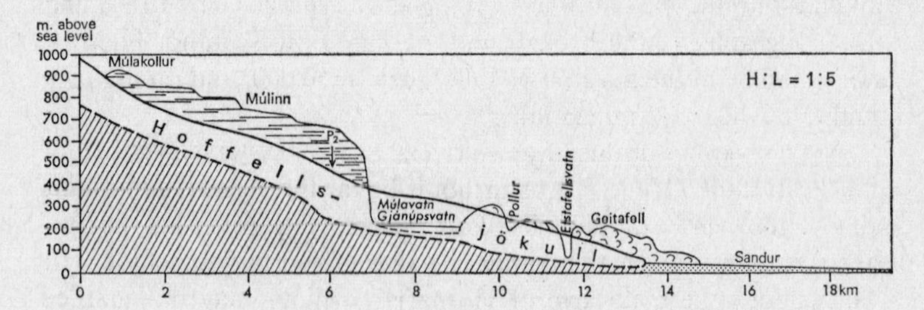

Fig. 6.2 Longitudinal profile of the lower part of Hoffellsjökull

from the southern outlet-glaciers of Vatnajökull. Austurfljót has one single point of origin, which is of the type found in almost every large sandur river. An enormous mass of water wells up almost vertically in front of the ice margin, which indicates that it is squeezed out under great pressure, and that the sandur plain in front of the glacier is considerably higher than the base of the glacier. I received confirmation of this in the spring of 1938 on Skeiðarársandur, where a recent 'jökulhlaup' had just broken up the ice front.

For the first 1500 m of its course, Austurfljót is confined to one channel. The slope of the sandur plain is here 1:60, and the river is very rapid. Afterwards it splits up, first into two arms, and then into more and more, thus creating a 'fan' of river arms. As in other sandur rivers, however, the bulk of the water is collected in a few (2–5) channels, which carry the greater part of the coarse material. In these distal parts of the sandur plain the slope is 1:110–1:140. About 6 km from its source the river runs into Hornafjördur.

Table 6.1 Total monthly ablation on Hoffellsjökull ($m^3 \times 10^6$) and mean monthly values of total meltwater run-off (m^3/s) for the years 1936, 1937 and 1938

	Total ablation ($m^3 \times 10^6$)			Mean monthly values of total meltwater run-off (m^3/s)		
Month	1936	1937	1938	1936	1937	1938
Jan	0·5	2·0	0·5	0·2	0·8	0·2
Feb	1·0	0·2	0·5	0·4	0·1	0·2
Mar	3·0	2·5	1·5	1·0	1·0	0·6
April	25·0	19·5	16·0	9·5	7·5	6·0
May	100·0	40·0	38·0	37·0	15·0	14·5
June	201·5	154·5	130·0	77·5	59·5	50·0
July	284·5	163·0	198·0	106·0	60·5	74·0
Aug	196·5	146·0	173·0	73·0	54·5	64·0
Sept	147·5	44·0	90·0	57·0	15·5	35·0
Oct	42·0	13·5	18·0	15·5	5·0	7·0
Nov	5·0	5·0	8·0	2·0	2·0	3·0
Dec	1·5	2·5	2·5	0·6	1·0	1·0
Year	1008	593	676	32·0	18·7	21·5

So far, no direct observations have been made of the volume of water carried by Hornafjardarfljót or any other Vatnajökull glacier river. Such observations would, as a matter of fact, be rather difficult owing to the varying courses and depths of the rivers and the percolation of water through the sand. Thanks to the ample material obtained by the Vatnajökull investigations, and primarily by the ablation measurements, however, the run-off from Hoffellsjökull through Hornafjardarfljót in the three years of investigations, 1936–8, can be estimated with fair certainty (Table 6.1). Curves

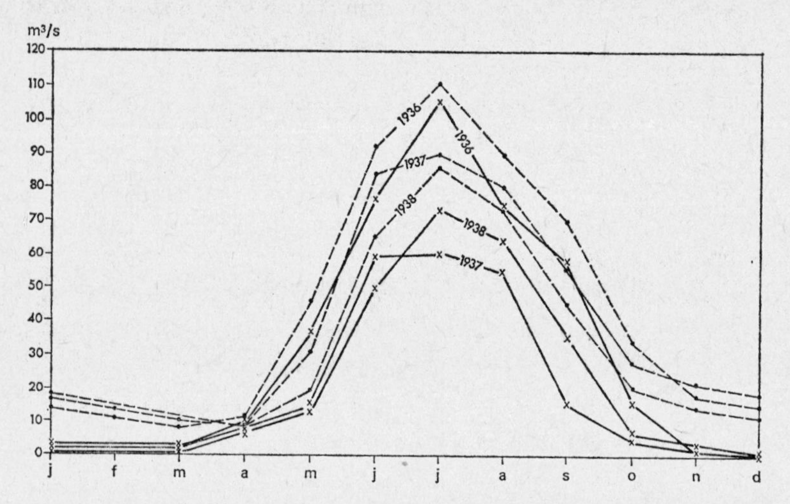

Fig. 6.3 The total ablation on Hoffellsjökull, expressed in average run-off per month (continuous curves), and the total average run-off per month of Hornafjardarfljót (broken curves) for each month of the years 1936, 1937 and 1938

showing the total run-off from the Hoffellsjökull district in 1936–8 are plotted in Fig. 6.3. These curves can naturally not claim to be exact, but they will nevertheless give a fairly correct idea, both quantitatively and qualitatively, of the drainage of the Hoffellsjökull district, which is typical of the whole southern part of Vatnajökull.

The annual régime of an average year includes a minimum in March–April of 5–10 m^3/s; the run-off is slightly increased towards the end of April or early in May, when ablation and precipitation in the form of rain increase. In June most of the snow on the higher levels melts, and the ice-dammed lakes at the eastern margin of the glacier, in which part of the meltwater from the late winter and spring has been stored, are drained, and the run-off therefore increases very rapidly. In July the ablation reaches its highest levels, and in most years also the run-off. As the influence of the ablation extends higher up in the firn area, the run-off is retarded. The run-off curve is pushed to the right in relation to that of the ablation. In September the precipitation in the form of rain reaches its maximum, and in this month the run-off is large in relation to the ablation. In October and November the run-off largely depends on the amount of rain, and on the time when the ice-dammed lakes are blocked up. Although ablation has practically ceased already in November and but little precipitation falls in the form of rain in the glacier area during the winter months, the river carries relatively large quantities of water right up to February, and the minimum does not occur until April–May. This indicates that internal drainage plays a fairly important part, especially in the upper parts of the glacier.

The largest monthly average run-off in the three years of investigation was some 110 m^3/s in July 1936. In the same month of 1937 and 1938 it was about 90 m^3/s. The total run-off varies less from year to year than the total ablation. The total ablation figures – 1008, 593 and 676 $m^3 \times 10^6$ in 1936, 1937 and 1938 respectively – correspond to total run-off figures of about 1300, 1150 and 1050 $m^3 \times 10^6$. This is quite natural, for heavy ablation, especially on the high levels, demands both high temperature and a large radiation income, and radiation is much reduced in the summer, when the precipitation is heavy. Two factors of great importance to the run-off – the summer rain and the ablation due to radiation – are thus counteracting one another.

THE SILT-CONTENT OF THE GLACIER RIVERS

Our 1936 working program included the taking of water samples from the Hoffellsjökull glacier rivers for the determination of their silt contents. We had accordingly brought a sample-bucket, designed by Statens Meteorologisk-Hydrografiska Anstalt in Stockholm, of exactly the type used and described by F. Hjulström (1935, p. 186). As other and more important work intervened, the samples taken were neither as many nor as systematic as would have been desirable. About twenty samples were taken, as far as possible at fixed hours and in the most suitable places.

Two samples were taken in Austurfljót after a day of exceptionally

Table 6.2 Contents of silt and dissolved salts in water

Sample No.	Date	Hr	Temp.* at Hólar (°C)	Precipitation at Hólar (mm)	Place where sample taken (Letters G–L refer to Fig. 6.1)
	1936				
1	24 June	10	15·3	0·0	*G* = Main arm of Austurfljót, 800 m from ice front
2	24 June	10	15·3	0·0	*H* = Main arm of Austurfljót, 2 m from ice front
3	10 July	14	12·2	0·7	*H*†
4	27 July	10	10·5	5·1	*H*†
5	27 July	14	10·5	5·1	*H*†
6	27 July	14	10·5	5·1	*G*
7	27 July	14	10·5	5·1	*G*
8	29 July	14	14·9	0·0	*I* = E arm of Vesturfljót, 200 m from ice front
9	29 July	14	14·9	0·0	*H*
10	6 Aug	14	11·7	0·0	*K* = E arm of Austurfljót, 800 m from ice front
11	10 Aug	10	13·4	0·0	*L* = One of main arms of Austurfljót, 4000 m from ice front
12	10 Aug	10	13·4	0·0	*L*
13	10 Aug	14	13·4	0·0	*L*
14	10 Aug	14	13·4	0·0	*L*
15	10 Aug	20	13·8	0·0	*L*
16	10 Aug	20	13·8	0·0	*L*
	1937				
17	30 July	14	11·7	84·5	*L*
18	30 July	14	11·7	84·5	*L*
19	30 July	14	11·7	84·5	} River Hoffellsá, between Hoffell and Setberg
20	30 July	14	11·7	84·5	}
	1938				
21	25 May	10	5·0	—	} Skeiðará, due W of Skaftafell, 4000 m from ice front
22	29 May	10	5·0	—	}

NOTES

* Temperature and precipitation are average values for the 24 hr preceding sampling.
† Samples taken close to the eddy where the river wells up at the ice front.
‡ The discharge of the whole Austurfljót.

(*continued opposite*)

samples from glacier rivers in Skaftafellssýsla, 1936–8

Depth of place (m)	Distance from bottom (m)	Approx. discharge (m³/s)	Silt content (mg/l)	Ignition loss (mg/l) and as % of silt content (mg/l)	Dissolved material (mg/l) and as % of silt content (mg/l)	
0·1	—	140	1463·9	36·1 (2·5)	—	
0·1	—	140	2000·0	42·4 (2·1)	—	
0·1	—	110	1355·5	32·3 (2·4)	—	
0·1	—	80	647·6	25·9 (3·9)	—	
0·1	—	90	822·4	37·0 (4·5)	—	
0·1	1·9	90	574·7	9·7 (1·6)	—	
1·8	0·2	90	597·3	22·1 (3·7)	253·1 (42·4)	
0·1	—	—	637·2	13·3 (2·0)	221·2 (34·7)	
0·1	—	130	1230·8	38·5 (3·1)	—	
0·1	—	—	223·9	15·7 (7·0)	—	
0·1	0·9	100‡	806·3	27·3 (3·4)	319·8 (39·7)	(a)
0·8	0·2	—	1081·8	40·8 (3·8)	—	(a)
0·1	1·0	110	847·7	18·8 (2·2)	—	(a)
0·9	0·2	110	1168·0	28·0 (2·4)	—	(a)
0·1	1·1	130	970·8	29·2 (3·0)	241·7 (24·9)	(a)
1·0	0·2	130	1238·5	32·1 (2·6)	—	(a)
0·1	1·6	300–400	1522·2	58·5 (3·8)	298·8 (19·6)	(b)
1·5	0·2	300–400	1541·1	58·9 (3·8)	241·4 (15·6)	(b)
0·1	—	> 100	745·4	27·3 (3·7)	50·0 (6·7)	(b)
1·0	—	> 100	787·5	33·3 (4·2)	54·3 (6·9)	(b)
0·1	> 8·0	> 50,000	14,920·0	188·8 (1·3)	348·0 (2·3)	(c)
0·1	—	> 2000	9610·0	77·2 (0·8)	304·0 (3·2)	(d)

(*a*) 10 August was a typical, rain-free summer day, medium warm and with normal cloudiness.
(*b*) Precipitation at Hoffell from 29 July, 14 h to 30 July, 14 h amounted to 140 mm.
(*c*) The 'jökulhlaup' in Skeiðará culminated on this day.
(*d*) The 'jökulhlaup' had almost run dry.

heavy rain (29–30 July, 1937), and at the same time two samples in the river Hoffellsá, which contains no glacier water. This was done in order to get some idea of the erosive effect of such rain.

In the spring of 1938 a 'jökulhlaup' occurred on Skeiðarársandur. Unfortunately I did not get there until it was over, but Ragnar Stefánsson, a Skaftafell farmer, had on his own initiative taken two water samples, one when the jökulhlaup was at its height (25 May), and the other four days later, and sent them both to me. All the water samples were analysed in Stockholm by Mr E. Karlsson of the Geological Survey of Sweden on the asbestos method (described by Hjulström (1935) pp. 394–7). The results are given in Table 6.2.

The samples from the main arms of Austurfljót (nos. 1–7, 9 and 11–18 in Table 6.2), which represent undiluted glacier water, are the most important. These sixteen samples contained on an average 1066·8 mg of silt per litre. Except for sample no. 10, which, being mixed with water from mountain rills, is not included in the calculation of the average silt content, that content is amazingly uniform. It varies regularly with the discharge, which could be approximately determined with the aid of the ablation data and the meteorological observations at Hólar. The largest silt content was found on 24 June, on which day the river carried most meltwater that summer. The samples taken on 27 July, when the discharge of the river was unusually small, contained least silt, and that content decreased rather rapidly from the ice front (samples 3 and 4) to a point only 800 m away from this (samples 5 and 6).

Samples 11–16 were all taken on the same day and in the same place, at L in Fig. 6.1, in one of the main arms of the river, 4500 m from the ice front. They show that the silt content increases regularly in proportion to the increase of the discharge and slightly, too, with the depth below the surface, and that this increase is fairly constant. The average silt content of these six samples was 1085·5 mg/l, almost the same as the average of all the samples taken in Austurfljót. This confirms that that average, 1066·8, or in round figures 1070 mg/l, *fairly accurately represents the average silt content of Austurfljót in the summer months.*

The average ignition loss of the Austurfljót samples was 33·6 mg/l or 3·2 per cent of the silt content. This is a relatively high figure for glacier water, and indicates some organic content. The reason is probably that Hoffellsjökull, like the other southern Vatnajökull outlet-glaciers in recent times, has passed over forest and marsh ground. This is confirmed by the quantities of peat and birch trunks brought to the sandurs by the rivers.

Some of the water samples were also analysed for dissolved material (Table 6.2). The Austurfljót samples proved to contain a fairly constant amount of such material, averaging 270·9 mg/l, or 28·4 per cent of their corresponding silt content.

The total amount of material carried by the main Hoffellsjökull river thus averages 1066·8 + 270·9 = 1337·7, *or in round figures* 1340 *mg/l.*

RATES OF DENUDATION

The rates of denudation of the Hoffellsjökull district can be approximately calculated from the values given.

In 1882 A. Helland attempted to calculate the rate of denudation of the whole Vatnajökull district. While travelling in Iceland in 1881 he sampled some of the rivers, among them Núpsvötn (silt content 570 mg/l), a rill in the middle of Skeiðarársandur (1509 mg/l), and Jökulsá á Breiðamerkursandi (1876 mg/l). From these analyses, and a computed precipitation on Vatnajökull of 1007 mm (= the precipitation observed at Berufjördur, less 25 per cent for evaporation), he calculated the annual degradation in the Vatnajökull district to be 0·647 mm per annum, or six times larger than on the Jostedalsbræen.

The Swedish–Icelandic Investigations indicate that the average precipitation on Hoffellsjökull is about 3500 mm per annum, with a run-off coefficient of 85 or 90 per cent. The annual run-off is thus about 3000 mm per unit area. If we further assume the quantity of material transported to be 1340 mg/l, this means that 4200 tons/km^2 are removed annually which, taking the loose material to be of a specific gravity of 1·5, *means the removal annually of a uniformly distributed layer* 2·8 *mm thick*.

For comparison it may be mentioned that, according to L. W. Collet (1925), the amount of suspended material transported from the Aletsch Glacier (la Massa) in 1913 and 1914 averaged 748 tons/km^2. The quantity transported from Hoffellsjökull is 3200 tons/km^2. A certain amount of ash from volcanic eruptions is of course embedded in Vatnajökull, but on Hoffellsjökull that quantity is insignificant. Sections dug in Hornafjördur prove that the total thickness of ash collected there in the last 1000 years does not average more than 5 cm. Assuming that the figure of 748 tons/km^2 from the

Aletsch Glacier is correct, the rate of denudation in the Hoffellsjökull district is at least four times greater. This is partly due to the very easily eroded nature of the rock in the Hoffellsjökull district, and partly to Hoffellsjökull's larger kinetic energy.

The rate of denudation computed from the silt observations, viz. 2·8 mm per annum, is of course a minimum value, as no allowance has been made for the material carried along the bottoms of the rivers. It is very difficult to say how large this may be in proportion to the suspended material. Collet (1916) found that, in the Rhône at Gampenen, the proportion is 1 : 1. *That* ratio is certainly too low for Hoffellsjökull, where the minimum degradation would thus be 5·5 mm per annum. *This means that a layer* 1 *m thick would be eroded in* 180 *years*. The amount of postglacial uplift in this district is 80–100 m. Probably, therefore, the rate of erosion has almost balanced the rate of uplift.

Samples 19 and 20 (Table 6.2) were taken in the river Hoffellsá, which drains the valley immediately east of Hoffell, in order to get some idea of the amount of denudation caused by violent rain in a non-glaciated part of Skaftafellssýsla. As late as in the 1920s this valley contained a branch of Lambatungnajökull, but this has now receded and the river no longer receives any glacier water. The samples were taken at 2 p.m. on 30 July 1937, when precipitation of 140 mm had been registered at Hoffell in the preceding 24 hours. The average precipitation in that time in the Hoffellsá drainage area above the point where the sample was taken had certainly been 200 mm.

The suspended material in the two samples averaged 766·4 mg/l, and the dissolved material, naturally a relatively small quantity, 30·3 mg/l. The total material transported was thus in round figures 800 mg/l. If we assume a run-off coefficient of 90 per cent, *this means that* 150 *tons/km*^{2} *of material was carried away in this one day, i.e. denudation averaging* 0·1 *mm*. For comparison it may be noted that the amount of denudation in the Hoffellsá district in this one day was nearly three times larger than that for one year in the Fyris area in Upland, Sweden, which, according to Hjulström (1935, p. 436), amounts to 0·037 mm. At normal water level, the silt content in Hoffellsá is much smaller, but the content of dissolved material is larger. *The annual amount of denudation in the non-glaciated parts of the Hornafjördur district may therefore be estimated at not less than* 0·5–1 *mm, only about one-fifth of that in the glaciated district*.

The two water samples taken from the Skeiðará jökulhlaup in 1938 (nos. 21 and 22 in Table 6.2) are the first to give us any idea of the amount of silt carried by such catastrophes. They averaged 12,600 mg of silt per litre. *If the total water volume of this jökulhlaup is estimated at about* 12 *km*^{3}, *the quantity of material removed from the Skeiðarárjökull district* (1722 *km*^{2}) *would, apart from material carried along the bottom of the river, amount to about* 150,000 *tons.*

REFERENCES

COLLET, L. W. (1916) 'Le charriage des alluvions dans certains cours d'eau de la Suisse', *Annales Suisses d'Hydrographie*, IX.

—— (1925) *Les Lacs* (Paris).

HELLAND, A. (1882) 'Om Islands Jökler og om Jökelelvenes Vandmaengde og Slamgehalt', *Arch. f. Mathem. og Naturv.*, VII, no. 1.

HJULSTRÖM, F. (1935) 'Studies of the morphological activity of rivers as illustrated by the River Fyris', *Bull. Geol. Inst. Uppsala*, XXV.

7 Radiating Valleys in Glaciated Lands

DAVID L. LINTON

AMONG the glacially-eroded landforms widely displayed in Scotland three have impressed me in the field as being particularly significant. First are the features known to continental workers as *cirques* or *karen*, but in Britain by the Scottish word *corrie*. Second are the *glacial troughs*, especially those in which true rock basins have been developed. The third group comprise those areas, mostly lowland or low plateau but in some cases including the flanks and even the summits of mountains, in which the preglacial character of the surface has been so transformed that we may speak of the present forms as being *ice-moulded*. Fields of *roches moutonnées* obviously fall in this category.

Now these three classes of glacial landforms are obviously associated with different degrees of glacierisation of the land mass. Corries held only small isolated bodies of ice and such corrie glaciers were certainly the first ice-masses to appear as glacierisation began in Scotland, and equally the last to survive as the climate ameliorated in interglacial and postglacial times. Their distribution, if plotted on a map, should tell us something of the climatic conditions of such initial or waning phases of glacierisation. Glacial troughs on the other hand, especially those of considerable size, evidently discharged large areas of ice – Highland Ice in the classification of C. S. Wright and R. Priestley (1922). Such areas may have been produced in some cases by the growth and confluence of corrie glaciers; more usually, we must suppose, a general deterioration of climate led to the appearance of extensive fields of *névé* beneath which many of the earlier-formed corries would be submerged. Finally, the ice-moulded forms imply the wholesale streaming of ice over the open country of the lowlands and low plateaux, reaching as far as the coastline and as high as the summits of such mountains as Suilven (713 m) in Sutherlandshire. The ice-body which gave rise to these forms must have invested the larger part of the country and the forms themselves thus belong to a phase of maximum glaciation: presumably, in the form in which we now see them they were fashioned by the ice of the last glaciation.

A general study of these three types of landforms in Scotland should therefore tell us something useful about the centres of ice accumulation and activity, and therefore of heavy Pleistocene precipitation, at these three contrasted stages of glacierisation. Such a study has recently been concluded by the present writer (1957). Its main conclusions are startlingly simple. The areas in which our three selected landforms are developed to the greatest extent, and hence the chief areas of glacial activity and of heaviest precipitation in Pleistocene times, turn out to be the areas of heaviest precipitation today (Fig. 7.1).

CORRIES

In the case of the corries, a glance at Fig. 7.2 will show that the overwhelming majority of the 473 corries mapped lie in the wetter west of the country, and show a close correspondence with the areas which are shaded on the map and today receive a yearly precipitation total of 2500 mm or more. Along the western seaboard from the Isle of Arran, through Argyll, Mull, western Inverness and Rum to Skye this relationship holds. In the north where summer temperatures were lower, and on the lee side of the mountains where a higher snowline led to the same result, a smaller yearly total appears to have sufficed and in these areas corries are found on mountains that today receive as little as 1800 mm precipitation a year. In the Cairngorms and eastern Grampians, the most continental part of the Highlands where the duration of snow cover is today at its maximum, it would appear that even less was sufficient, for corries are found on mountains receiving only 1500 mm precipitation a year today.

ICE-MOULDED FORMS

If we turn from the incipient and final stages of glacial activity to its maximum phase as represented by the ice-moulded landforms a similar contrast between east and west is apparent. With little exaggeration it may be said that along the whole west coast from Cape Wrath to the Mull of Galloway, except where mountains reach the sea (as in Torridon, Skye, Morvern or Mull), or the rock surface of the lowlands is masked by drift (as in the Rhinns of Galloway), ice-moulded forms are universal. In the Lewisian Gneiss country of Sutherland and Wester Ross they are strongly developed and take a

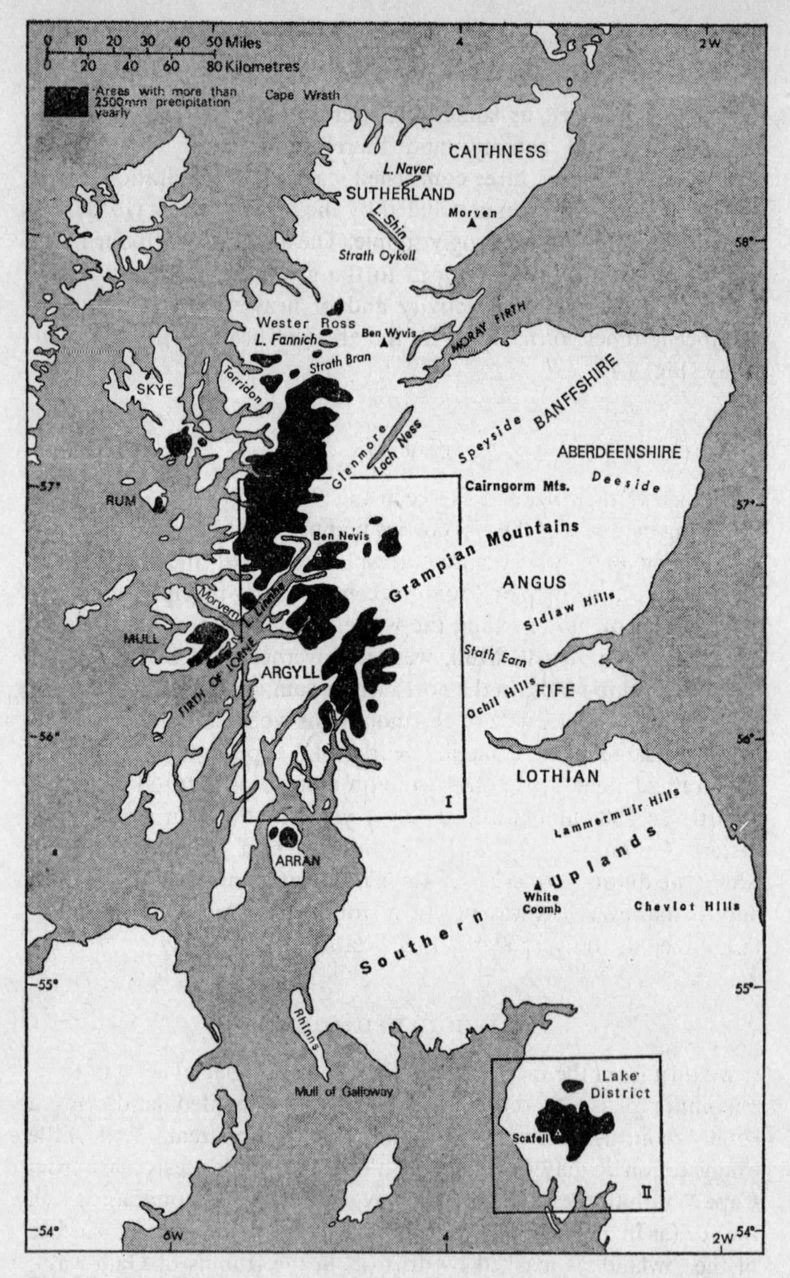

Fig. 7.1 General map of northern Britain showing names used in text, all areas receiving more than 2500 mm annual precipitation and the areas covered by Figs. 7.3 and 7.5

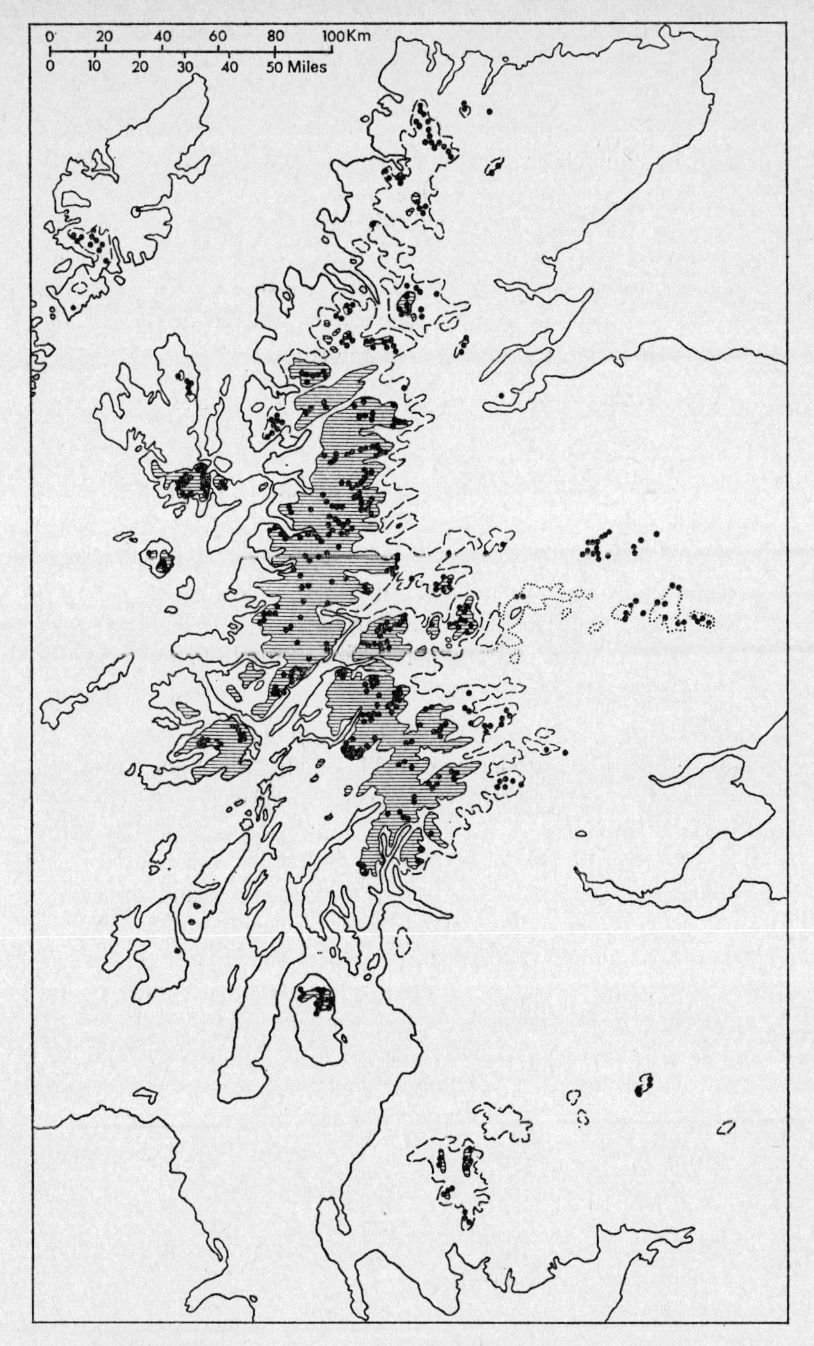

Fig. 7.2 Pleistocene corries and present-day precipitation in Scotland. Each black dot represents one corrie. Shaded area bounded by full line receives more than 2500 mm annual precipitation. Dashed line = isohyet for 2000 mm. Dotted line = isohyet for 1500 mm

form which may be described as *knoek and lochan* country; in western Argyllshire against the Firth of Lorne the coincidence of rock strike with the main direction of ice motion has produced a strikingly ridged and grooved landscape of *rock-drumlins*.

By contrast, these forms practically never reach the coast and the exceptions to this rule are most instructive. Ice-moulded forms are found in the east only where powerful ice-streams from western sources are known from other evidence to have descended toward the North Sea. These taken in order from north to south are as follows:

(1) Helmsdale, Strath Brora and the Kyle of Sutherland.
(2) The glens that open to Cromarty Firth on either side of Ben Wyvis.
(3) The low plateau and hill country on both sides of Loch Ness
(4) Speyside.
(5) Deeside.
(6) Strath Tay and the hills of Angus.
(7) Strath Earn, the southern Sidlaws and eastern Ochils.
(8) The Forth lowland, southern Fife and the Lothians.
(9) The Teviot–Lower Tweed lowland.

Between these avenues of major ice-streaming were uplands upon which the ice appears to have exerted little effect and upon which the writer has noted the survival of essentially preglacial forms (Linton, 1949*a*). The existence of remnants of former base-levelled surfaces surmounted by monadnocks, of polycyclic valley forms of normal erosion, of tors (Linton, 1955) and, in places, of a mantle of deeply-rotted rock sometimes preserved beneath glacial deposits, points clearly to a minimum of glacial erosion on the upland portions of eastern Scotland from Morven in Caithness, through Banffshire and Aberdeenshire, to Lammermuir and the Cheviot. Glacial erosion, even at the stage of maximum glaciation, is found to be very unequally distributed over Scotland, and the differences between east and west amount to the difference between extremely severe erosion with complete re-moulding of the surface, to possibly none at all.

GLACIAL TROUGHS IN BRITAIN

We turn lastly to the glacial troughs, believing that they were chiefly used during phases when the land carried much more ice than in the corrie glaciations and much less than in the glacial maxima. In this respect the troughs of the south-western Highlands appear to be particularly instructive (Fig. 7.3). A radiating series of some fifteen major glacial troughs, not counting diffluent branches, may be enumerated. It is clear that the system has accommodated itself in several instances to the existence of ready-made depressions following the northeast–southwest Caledonian strike of the Dalradian rocks of the area. It is clear too that the system is asymmetrical, being extensively developed toward the southwest, south and southeast and not at all toward the north and northwest, since escape was possible in the former directions but not in the latter which lead toward other high mountains acting as centres of ice accumulation.

This radiating system must have been fed by a major centre of ice accumulation at its heart. Today the mountain masses of this area receive over 2500 mm annual precipitation, and if we draw a line to enclose all these areas and so depict a generalised area of very heavy present precipitation we find that, except on the east, the lakes and fiords lie on its margin. On the eastern side they lie well beyond it (Loch Tay 12 km), but possibly here, as with the corries, a lower isohyet should be selected on the lee side of the ice-shed. The 2000 mm isohyet in fact intersects Loch Rannoch, Loch Tay, Loch Earn, Loch Lubnaig, Loch Achray and Loch Ard.

We thus define an ice-dome which was possibly the most powerful centre of ice-dispersal in Britain. The ice from it overran and largely remodelled the surface of the Lorne plateau to the southwest. It branched in the south into diffluent streams too numerous to be separately identified without further investigation, for in the region between the heads of Loch Fyne and Loch Lomond no watershed at the head of any valley has survived intact. Farther east, diffluent ice spilled eastward out of Loch Lomond at two or more points; ice from Loch Voil made a permanent breach in the Forth–Earn watershed now occupied by Loch Lubnaig (Linton, 1940) and the Tay ice spilled across Glen Ogle into Strath Earn. Most impressive of all the evidence of the power of this ice-mass are the clean breaches in the main east–west Grampian watershed now occupied by Loch Ericht and Loch Treig. I have elsewhere estimated the extent of lowering accomplished by the ice in these two cases at 350 and 550 m (Linton, 1949*b*). It is clear that the dispersive tendency of a major ice centre of this kind is capable of imposing a system of radial outlet troughs on the preglacial landscape even if the valley pattern of the latter is only partly favourable to the development of such a system. The trough

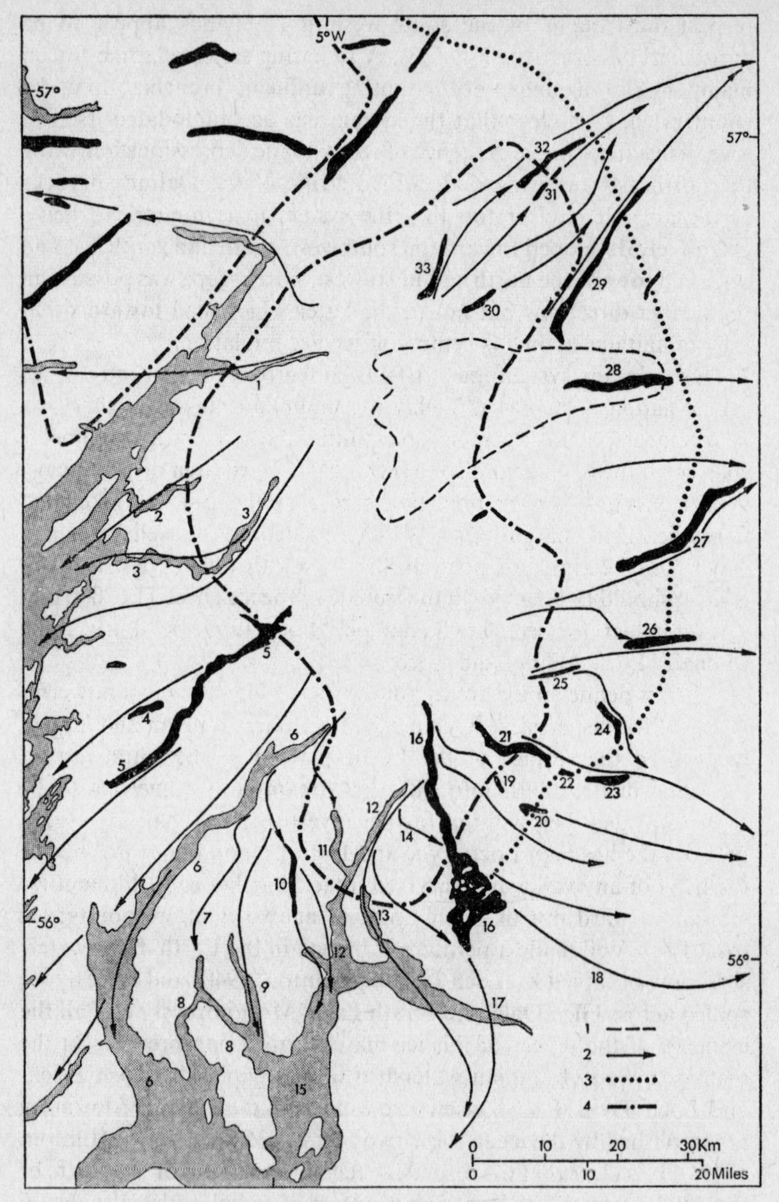

Fig. 7.3 Radiating troughs and present-day precipitation in the south-western Highlands of Scotland

system will use and enlarge those valleys (such as Loch Awe, Loch Tay and Loch Earn) which it finds lying in the right direction and ready for use, but where such are not available it will adapt the existing valley pattern by overriding watersheds and linking existing valley elements, or even creating entirely new ones such as those containing Loch Lubnaig, Loch Ericht and Loch Treig. Wherever this has occurred it will obviously be difficult to recognise the pattern of the drainage system before glacial modification took place, and misleading conclusions will be drawn if the whole radiating system is taken to be preglacial. In another British area the radial valley system has been universally accepted at face value and in consequence an erroneous view of its origin and of the tectonic history of the region has been held for a century. This case, that of the English Lake District, merits special attention.

The wheel-like pattern of the valleys of the Lake District was recognised and commented on by the poet Wordsworth, and was explained by the mathematician and geologist William Hopkins as early as 1848 as resulting from doming of the rocks. Later workers, particularly J. G. Goodchild (1888–9) and J. E. Marr (1906), accepted this general view and pictured the drainage of the area as originating on a surface of Cretaceous rocks uplifted into a local dome during Tertiary times, and later becoming superimposed upon the Palaeozoic

---

Caption for Fig. 7.3 (continued)

Key to line-types

1. Boundary of the enclosed depression of the Moor of Rannoch
2. Stream-lines of ice movement
3. Generalised present-day isohyet of 2000 mm
4. Generalised present-day isohyet of 2500 mm

Glacial Troughs

1. L. Leven
2. L. Creran
3. L. Etive
4. L. Avich
5. L. Awe
6. L. Fyne
7. Glendaruel
8. Kyles of Bute
9. L. Striven
10. L. Eck
11. L. Goil
12. L. Long
13. Gare Loch
14. Glen Douglas
15. Firth of Clyde
16. Loch Lomond
17. Vale of Leven
18. Strath Blane
19. L. Chon
20. L. Ard
21. L. Katrine
22. L. Achray
23. L. Vennachar
24. L. Lubnaig
25. L. Voil
26. L. Earn
27. L. Tay
28. L. Rannoch
29. L. Ericht
30. L. Ossian
31. Lochan na Earba
32. L. Laggan
33. L. Treig

rocks in which the valleys are now wholly entrenched. They recognised further, as indeed had Hopkins, that the uplift could not have been a perfectly symmetrical dome. Although showing true radial symmetry at its western end, it has only axial symmetry in the east,

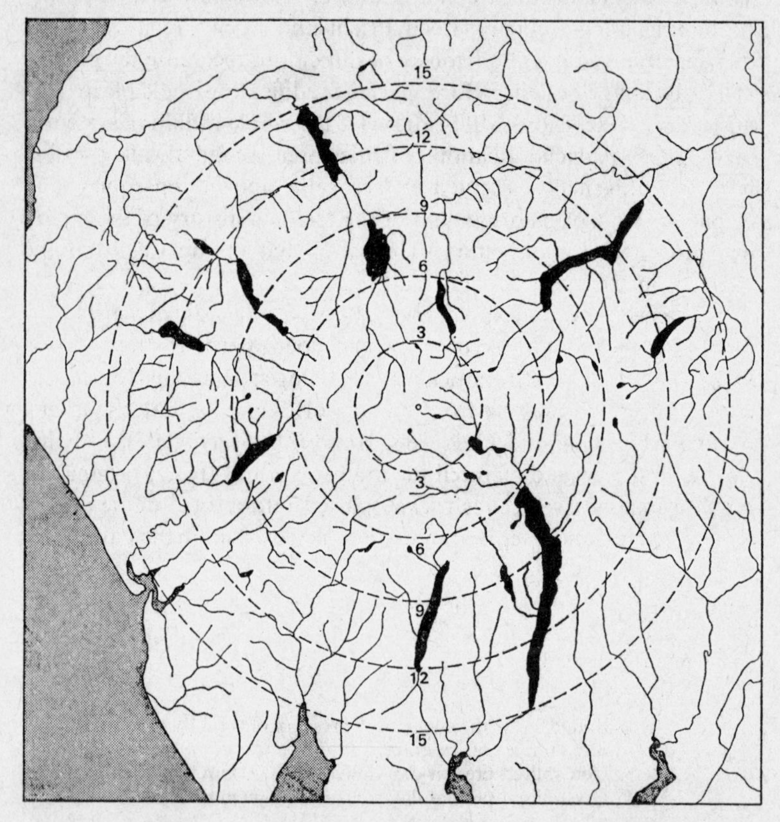

Fig. 7.4 Map drawn by B. V. Darbyshire and used by H. R. Mill in 1895 to illustrate the radiate symmetry of the Lake District

the axis of uplift running from Scafell by the Kirkstone Pass toward the east or east-southeast.

If, however, we examine the radial symmetry more closely it will be found to be a symmetry of the valleys containing lakes and not of the drainage pattern as a whole. This was clearly displayed in a map drawn by B. V. Darbyshire (Fig. 7.4) and used by H. R. Mill to illustrate the account of his bathymetrical survey of the lakes published in 1895. On that map, circles of 3, 6, 9, 12 and 15 miles radius were drawn about a centre of symmetry located at High Raise (National Grid Reference NT 290090). Except for the smaller lakes Thirlmere and Grasmere the inner area is lake free, and the circle of 6 miles radius 'may be taken as the commencement of the radiating lake system'. The circle of 9 miles radius 'may be termed the central line of the Lake District, passing, as it does, through or near the deepest part of the four deepest, and across the alluvial flats which separate the upper and lower members of the two lakes which were once single, but are now divided'. Only six lakes project beyond the 12-mile circle and the 15-mile circle includes them all. In a note at the end of this paper E. Heawood remarks on the way 'the lakes as a whole reach just as far as and no further than the beginning of the more level country that skirts the district'.

Mill and Heawood were not primarily concerned to explain the facts they had so diligently laboured to make available to science, but it is clear now that they did in fact demonstrate that the radial symmetry is a feature of the glacially-scoured rock basins and not of the valley system as a whole. This may be made quite clear by consideration of two wholly unrelated sets of facts.

(1) As in the southwest Highlands the region within the area of radiating valleys is one which today has excessive precipitation ranging up to more than 4000 mm a year. If the generalised isohyet of 2500 mm be drawn as before it will be found that the troughs of Haweswater, Ullswater, Derwentwater–Bassenthwaite, Buttermere–Crummock Water, Ennerdale Water and Wastwater all lie radially upon the margin of the area of high precipitation so enclosed (Fig. 7.5). Only Coniston and Windermere lie beyond this line and indeed, if we generalise our isohyet sensibly and rectilinearly on the south even this is not true. There can be little doubt here, just as in the southwest Highlands, that we are outlining an area of very active ice dispersal for which the lake-holding valleys were the main troughs of discharge. These may therefore have been modified by the ice from the preglacial valley pattern and their radiation is not necessarily an original feature of the drainage.

(2) It is a striking fact that the axial symmetry about a roughly east–west axis noted as characterising the eastern part of the Lake District continues beyond the confines of the Lake District both to east and west. Without entering here on any discussion of the precise

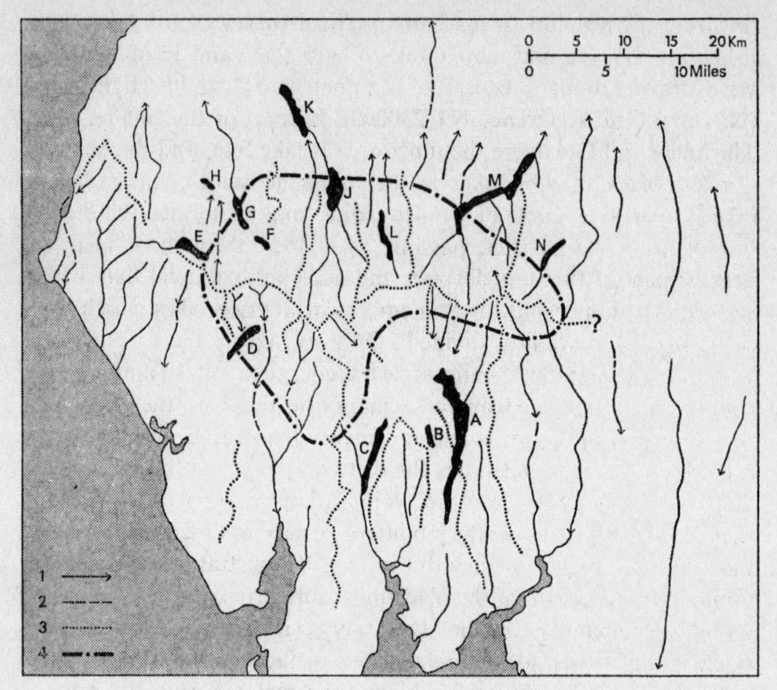

Fig. 7.5 Radiating troughs, present-day precipitation and some elements of the preglacial drainage in the English Lake District

Key to line-types

1. Surviving elements of early north-flowing and south-flowing drainage
2. Reconstructed elements of this system
3. Surviving watersheds related to the early north-flowing and south-flowing drainage
4. Area in which mountains today receive more than 2500 mm precipitation a year

Lakes

A Windermere	E Ennerdale Water	J Derwentwater
B Esthwaite Water	F Buttermere	K Bassenthwaite
C Coniston Water	G Crummock Water	M Ullswater
D Wastwater	H Loweswater	N Haweswater

position of the original water-parting in its eastward extension we may simply note that in this area the parallel north-flowing Lowther, Leith, Lyvennet and Hoff Beck tributaries of the Eden are balanced by the south-flowing Kent and Lune systems. In the north of the Lake District several of the streams have northward courses while in the Furness district of the south a north–south direction is the rule. In the low ground of the west the water-parting between the north-flowing Cocker and Marron, and the south-flowing Ehen and Keekle can be traced to the coast which it reaches midway between Whitehaven and Workington. In other words there is a good deal of evidence that the original drainage of the country was not radial but ran northward and southward from a watershed that can be traced across more than 60 km of country with a bearing about 282° and passing approximately through the well-known pass of Dunmail Raise which still separates the two drainage systems.

Thus, it is suggested that the Lake District ice-dome grew up in an area whose original valley system offered fairly free egress to north and south, and that such adjustment to structure as had already occurred had opened up some outlets to the low ground to the west; this pattern the ice was able to modify in the direction of a system of radiating outlet troughs which I believe to be natural to such ice-domes.

RADIATING VALLEYS IN SCOTLAND, NEW ZEALAND AND NORWAY

The Lake District ice-dome would appear to have measured about 40 by 25 km and to have had a solid foundation of high mountains (800–1000 m) at its centre. The ice-dome of the southwest Highlands was larger, about 100 by 50 km, and contained within it the large basin-like area of Rannoch Moor. The latter is about 25 by 12 km and some 300 m above sea level but 600–800 m below the summits of the surrounding mountains save on the east. Yet there is little doubt that it was filled up to mountain level by the ice and that ultimately the ice of the dome lay thickest and probably highest over the basin area.

To the north of the great dividing line of Glenmore the northwest Highlands supported another and even larger area of ice accumulation and dispersal. Doubtless it was initiated over the excessively wet mountains that lie between Loch Linnhe and the 57th parallel. Here for a tract 100 km long the annual precipitation is continuously over 2500 mm and generally over 3000 mm. Farther north the modern precipitation lessens as the mountains become lower, more broken and interrupted by basin areas holding, from south to north, Strath Bran and Loch Fannich; Strath Oykell and Loch Shin; and Loch Naver. There is good reason to think that in time these depressions, like the Moor of Rannoch, filled with ice and became part of a major ice-dome covering virtually the whole north of Scotland, 200 km long and 50 km broad. Some of this ice discharged to the Moray Firth by openings between the coastal hills from Morven to Ben Wyvis but no large rock basins were excavated, though small ones like Loch Luichart, Loch Glass, and Loch Morie are found. Some of the ice escaped to the north coast scouring out the basins of Loch Loyal and Loch Hope. But, since the ice-shed lay well eastward of the main mountain area, most of the ice passed through or over the mountains, breaching the main watershed to create innumerable low-level passes that lead to the heads of a great series of fiords and fresh-water lochs. In the north these lochs naturally open north-westward, those in the centre and south, westward and south-westward. The power of the ice that excavated them may be judged by the fact that the floor of the fresh-water Loch Morar descends 300 m below sea level, while in the Inner Sound where the ice was deflected by the Isle of Skye sounding has revealed a submarine basin 320 m deep.

To find examples of still larger systems of radiating glaciated troughs we must look outside Britain. Two areas seem to me particularly interesting. One of these, in New Zealand, I know only from small-scale maps and the descriptions of others (Fig. 7.6); the second, southern Norway, I have studied in the field (Fig. 7.7).

The district known as Fiordland in South Island, New Zealand, is a mountain block structurally distinct from the Southern Alps, cut off by an abrupt and probably recently faulted coastline on the west and by the tectonic depression of the Waiau valley and Lake Te Anau on the east. Its summits rise from 1200 m in the south to 2800 m in the north and it measures some 200 km from north to south and about 70 km across. Most of the area receives well over 2500 mm precipitation a year rising to 5000 mm in the western coastal mountains, but falling off sharply on the lee side to only 1700 mm even before the eastern border of the upland is reached. The modern distribution of precipitation is thus markedly asymmetrical falling from west to east, though the summit heights tend to rise in that direction (W. N. Benson, 1935). The area, however, exhibits a markedly symmetrical series of radiating troughs. Those of the northwest coast from Milford Sound in the extreme north southward to Daggs Sound offer a spectacular series of fiords 12–33 km long and, except for Bradshaw Sound which trends west, directed northwest. On the eastern side of the upland in the same

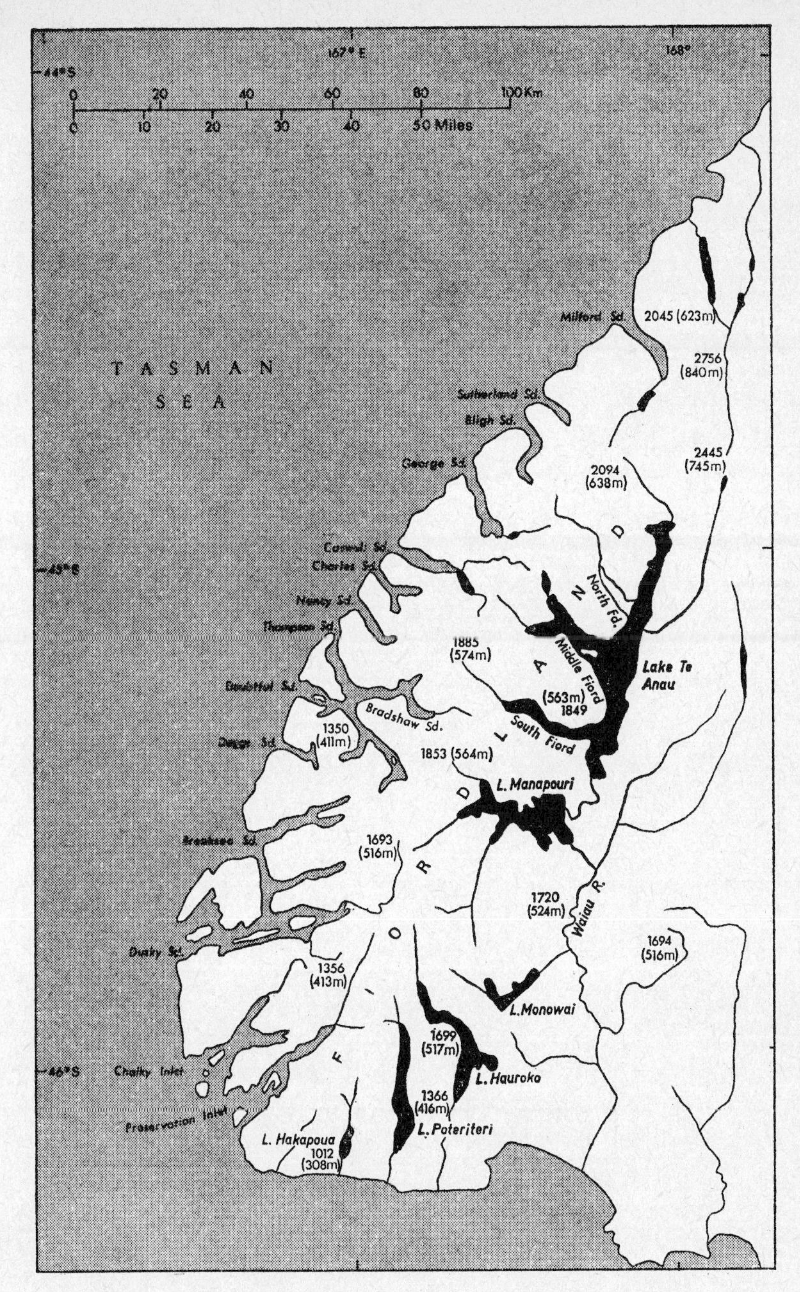

Fig. 7.6 Fiordland District, South Island, New Zealand, showing names used in text and characteristic summit altitudes

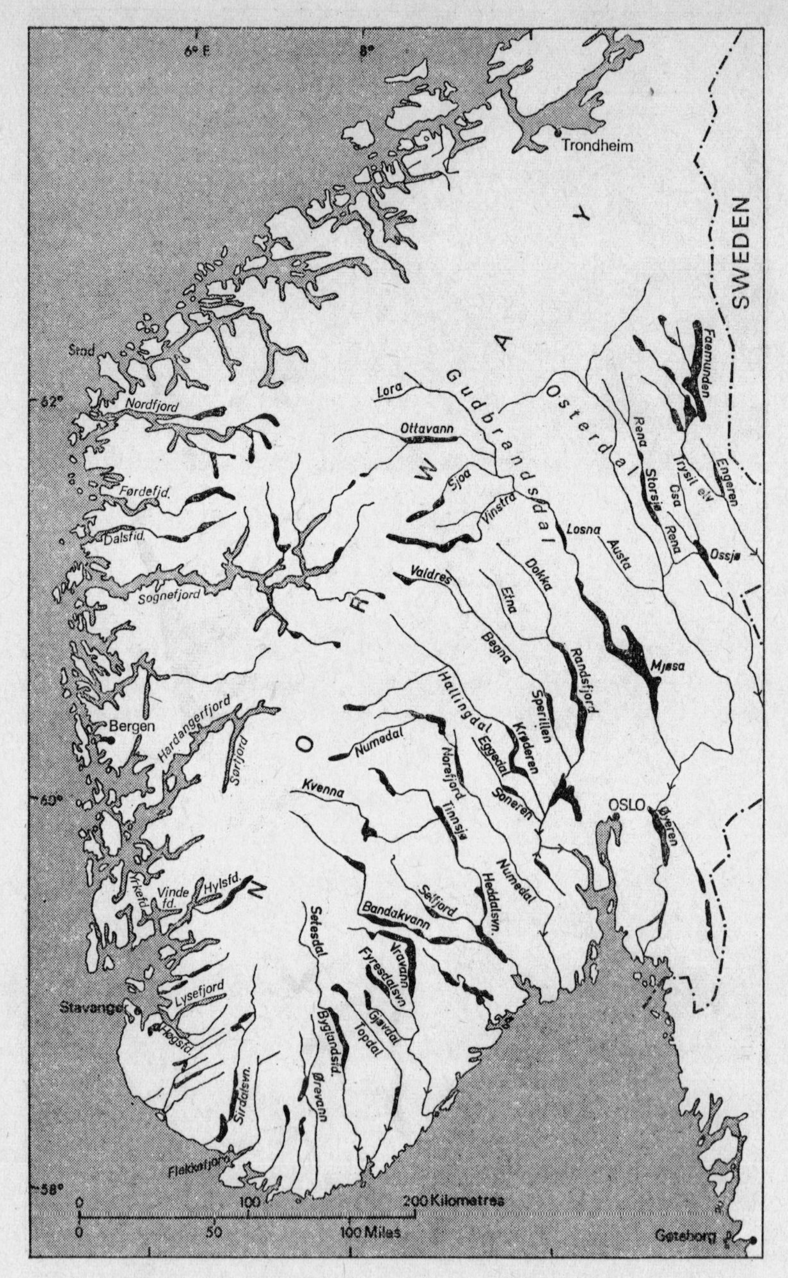

Fig. 7.7 Southern Norway showing names referred to in text

latitudes the North, Middle and South Fiords are branches of Lake Te Anau occupying troughs that in the first two cases trend southeast. These, like the northwesterly fiords of the northwest coast, may owe much to structural guidance. South Fiord trends east and continues the line of Bradshaw Sound. South of this line the sounds of the west coast trend west-southwest and then southwest, the lakes (Manapouri, Monowai, Hauroko, Poteriteri and Hakapoua) of the southeast and south swing round from east through southeast and south to south-southwest. The trough system as a whole is remarkable in the degree to which it is below water (much more so than in the area of the northern Highlands of Scotland which is of comparable size), and is of interest in that, in the north, the ice appears to have found structurally-guided parallel valleys that opened in suitable directions on both western and eastern margins, and in that the area of ice shedding again lies well to the lee of the zone in which precipitation is heaviest today.

The upland of southern Norway is a much larger area, measuring some 650 km along the height of land which runs roughly from south of Trondheim to Flekkefjord on the south coast, and about 350 km at right angles to this line. Glaciated troughs radiate from the upland in all directions except north, northeast and east which were of course precluded by the pressure of Swedish ice. These troughs fall into four series. The first comprises a great series of parallel troughs which begins near the Swedish frontier and extends west-southwest for 250 km. Their uniformity of direction is remarkable, their bearings ranging only between 136° and 164°, and they carry a striking series of large trough lakes, of which two bear the name fjord as though they belonged to the west coast. These parallel troughs may of course represent an original consequent drainage system. But the existence of abrupt elbows in many of the valleys with a turn from a generally easterly direction to the direction of the trough system; the existence of through depressions leading eastward from valley to valley, and of important west–east drainage elements in the higher parts of the upland (Etna, Valdres, Vinstra, Otta, Lora, etc.) all suggest that the preglacial drainage contained important west–east elements.

The second series occupies the south and is markedly radiate with a centre of radiation on the high ground of northernmost Aust Agder. The troughs in the eastern quadrant range in bearing from 109° through south to 197°, those on the western slope bear from

236°–250° and the influence of the 'Caledonian' strike 236°–238° in Rogaland is obviously marked. But while many troughs, as in Argyllshire, follow directions earlier taken by structurally guided valleys, it would appear that the Lysefjord trough is continued west of Högsfjord by a depression opened crudely across the grain of the country, and the composite trough that holds Hylsfjord, Vindefjord and Yrkefjord is broken across the course of depressions and fiords trending south-southwest or south-southeast.

Beyond Stavanger the great fiords of the west coast form a third series on a scale altogether larger. They have the expectable general westerly directions but are not so much new troughs as whole drainage systems up to 200 km in length, occupied and adapted by the ice till the original valleys have been deepened and transformed beyond recognition. These systems are much complicated by troughs at strikingly divergent angles (Sörfjord in Hardanger, 10°) which have made use of valleys following lines of structural weakness in the Bergen arcs and elsewhere. The fourth series comprises those of the northwest coast from the promontory of Stad to Trondheim. Long leads parallel to the coast and continued by longitudinal depressions on land are the feature here and evidently embody preglacial adjustments to structure. At right angles to these are troughs directed radially north-northwest. Nothing like all of these would be needed in an integrated drainage dominated by subsequent streams and we must conclude that many of them are 'short-cuts' opened up by the ice to secure direct radial passage.

This radiation is impressive and I believe that it owes more to glacial erosion and less to the preglacial stream pattern than the conservative might be disposed to concede. But unlike the small-scale radiations of the Scottish Highlands and the English Lake District it bears no relation whatever to the present distribution of precipitation. The great fiords of the west coast move up the present-day precipitation gradient and the wettest parts of southern Norway are all close to the west coast (Alfotbre 5000 mm, Nordhordland 4000 mm, Folgefonn 5000 mm). Moreover, the great parallel valley system noted above occurs in a part of the country where only the higher mountains today receive as much as 1000 mm and the valleys are very dry. Even in northern Scotland there is evidence that the ice-divide moved eastward away from the area of highest precipitation as glacierisation of the country became complete. In Fiordland in New Zealand it is necessary to assume a similar development on a rather larger scale to account for the observed radiation of the glacial troughs. In southern Norway at a still larger scale the divorce between the modern locus of high precipitation and the axis of the ice-cap at maximum glaciation is practically complete. Whereas the former lies everywhere within 30 or 40 km of the west coast, the latter axis slants obliquely across the country through Rondane, Jotunheimen, Hardangervidda to the neighbourhood of Storefonn and Breifonn. It is perhaps significant that these are the locations of the most easterly ice-bodies found in southern Norway today.

CONCLUSION

The conclusion of this discussion may perhaps be put in the following rather general terms. The ice of the smaller sorts of ice-bodies, corrie glaciers and glaciers of alpine type, moves downhill under the influence of gravity along paths which are almost wholly determined by the pre-existing topography, and the erosive work they do can modify only the style and not the ground plan of the landscape. On the other hand the ice of an ice-body large enough to cover the whole of the land mass which nourished it and to reach the open sea will also descend under gravity but in a simple radial outward movement *en masse*, unimpeded by the topography submerged beneath it. The ice cap that covered Iceland at periods of glacial maxima was doubtless of this form, as was the ice that streamed westward from the Scottish mainland to the Atlantic. This ice has refashioned the topography of almost the whole chain of the Outer Hebrides, yet had those islands not been there we should have known little of the extent and power of the ice that traversed them.

Between these two extremes are ice-bodies whose central parts are ice-domes wholly submerging the uplands that gave rise to them but whose marginal portions are constrained to some degree by the pre-existing topography. The tendency for free radial outflow exists powerfully but it is canalised by the relief of the land along favourably disposed preglacial valleys, or along new troughs produced by the integration of pre-existing valley elements by breaching of watersheds. Such ice-domes may thus do much to impart a radiate pattern to the valley system of a glaciated region and care must be taken not to attribute valley patterns of this origin to drainage initiation by doming or similar causes. Such ice-domes naturally range in size. The smaller examples such as those of the English Lake District or the southwest Highlands of Scotland are largely coincident with the region of greatest present as well as Pleistocene precipitation. The larger examples grew to cover areas of both mountain and basin topography and, as in southern Norway, spread far to leeward of the original areas of maximum snowfall. The smaller domes may be safely inferred from the coincidence of radiate troughs and an area of excessive modern precipitation. The larger reveal their former existence by the coincidence of a radiate trough system with the evidence of ice dispersal provided by travel of erratic stones.

REFERENCES

BENSON, W. N. (1935) 'Notes on the geographical features of south-western New Zealand', *Geogr. J.*, LXXXVI.

GOODCHILD, J. G. (1888–9) 'The history of the Eden and some rivers adjacent', *Trans. Cumberland and Westmorland Assocn.*, XIV.

HOPKINS, W. (1848) 'On the elevation and denudation of the district of the Lakes of Cumberland and Westmorland', *Q. J. geol. Soc. Lond.*, IV 70.

LINTON, D. L. (1940) 'Some aspects of the evolution of the rivers Earn and Tay', *Scott. geogr. Mag.*, LVI esp. p. 74.

—— (1949*a*) 'Unglaciated areas in Scandinavia and Great Britain', *Irish Geogr.*, II.

—— (1949*b*) 'Watershed breaching by ice in Scotland', *Trans. Inst. Br. Geogr.*, XVII esp. p. 14.

—— (1955) 'The problem of tors', *Geogr. J.*, CXXI.

—— (1957) *Morphological Contrasts of Eastern and Western Scotland.* Memorial Volume to A. G. Ogilvie (Edinburgh: Nelson).

MARR, J. E. (1906) 'The influence of the geological structure of the English Lakeland upon its present features: A study in physiography', *Q. J. geol. Soc. Lond.*, LXII.

MILL, H. R. (1895) 'Bathymetrical survey of the English Lakes', *Geogr. J.*, VI 46–73, 135–66.

WRIGHT, C. S., and PRIESTLEY, R. (1922) 'Glaciology', in *Results of British (Terra Nova) Antarctic Expedition of 1910–1913* (London: Harrison).

8 The Forms of Glacial Erosion

DAVID L. LINTON

ICE-MOULDED FORMS

GLACIAL erosion is universally attributed to two processes – abrasion and plucking – and though satisfying explanations in physical terms have yet to be offered of the mechanics of either, their reality and their general characteristics are not in doubt. Abrasion is self-evidently witnessed by polished and striated surfaces, by fluting and streamlining and by such distinctively original forms of ice-moulding as *roches moutonnées*. Plucking is evidenced in a variety of ways. There is G. K. Gilbert's (1910) argument that in southern Alaska where there is clearly evidence 'of an enormous amount of glacial degradation, it was a matter of surprise to find the reduction of the surface to smooth sweeping curves was a somewhat rare phenomenon. By far the greater number of exposed glaciated areas, even where degradation has been profound, abound in low embossments and in more or less angular groins or re-entrant spaces showing little trace of abrasive action.' This is equally true elsewhere. There is J. P. Dana's argument (cited in Gilbert, 1910, p. 207) that the blocks and boulders in the moraines of the Laurentide ice sheet, which received no falling stones from mountain peaks, must have been plucked bodily from the underlying bed. In Britain this argument is best exemplified by the blocks of Silurian grit that have been torn from the floor of Crummack Dale and carried uphill to rest on the limestone scar that overlooks the Craven fault. But plucking has its most conspicuous results in the cliffs of corries, though as late as 1896 E. Richter thought that because such cliffs show no signs of abrasion they were not glacially sculptured but fashioned by frost weathering above the surface of the ice. Today it would probably be accepted that both plucking and abrasion are involved in glacial erosion in most situations. In the surfaces described by Gilbert the embossments are rounded by abrasion and the groins reflect recent plucking, but both processes go on *pari passu* and even on the plucked lee sides of *roches moutonnées* smoothing is in evidence. Plucking by itself may fashion corrie cliffs throughout most of their extent but even here W. V. Lewis has commented on the abrasion of the lower portions.

As was noted by Gilbert, plucking is more in evidence on hard rock outcrops than soft, and the Borrowdale Volcanics of the English Lake District furnish an obvious example of the quarrying of large blocks from a massive rock, while the tapered and streamlined forms that I have described from Strath Allan and Strath Earn (Linton, 1962) seem good cases of the abrasion of soft sandstones and marls. Yet we have no evidence that the plucking process, however it operates, is not effective in removing blocks whose dimensions are measured in inches or a foot or two, and it is clear that nothing larger can be produced from close jointed or fissile rocks however hard they may be intrinsically. Conspicuous plucking is related to widely spaced jointing. Selective plucking along fault-lines, joints, shatter-belts and dykes in an area of very hard rock is probably responsible for the distinctively ice-moulded landscape of the Lewisian gneiss tracts of north-west Scotland. In this landscape, which I have elsewhere termed 'knock-and-lochan' topography, it is the depressions with their angular junctions, elbows, and parallel alignments that are the master features, though the rounding of the intervening hills ('knocks') by abrasion is ubiquitous. Similar topography is found again on the plateaux of well-jointed Lower Old Red Sandstone lavas of Lorne, but a little farther south against the Sound of Jura, where the outcrops of the Dalradian schists and quartzites run accordantly with the flow-lines of the ice, conspicuous streamlined hills are developed on the harder outcrops. To these the name 'rock drumlins' given by H. L. Fairchild in 1907 to features of similar form in New York State may fairly be applied, even though Fairchild's 'rocdrumlins' were cut in soft level-bedded shale and lightly plastered with till. Features homologous with the Argyllshire rock drumlins but of larger size – up to 5 miles in length – occur as a swarm in the Icelandic parish of Hreppar in the inner part of the south-west lowland, and again owe their strong development to the concordance of the directions of ice flow and underlying structure. In northern Iceland, however, in the lower valleys of the Blandá and Viðidalsá long rock drumlins in the two valleys curve convergently as the bay of Húnaflói is approached and their directions can reflect only the streamlines of the ice.

Ice moulding is not confined to lowlands, as the lake-studded interior plateaux of Iceland or southern Norway, or the knock-and-lochan topography on the uplands on either side of Loch Morar in the west Highlands, all testify. In the last case the upland is nearly

2000 ft above the lake, and the latter is just over 1000 ft deep, and the interesting question arises how thick was the ice that streamed over the upland to produce irregularities with an amplitude of some scores of feet. One feels that many hundreds of feet would be necessary for the purpose, and two pieces of quite extraneous evidence support this view. One is Fairchild's estimate – which he was careful to describe as 'merely suggestive' although it is very reasonably argued – that the ice was some 900 ft thick over west central New York and more than 700 ft of ice covered the largest drumlins. The other is afforded by one of the newly published sheets (1/10,000) of the Aletsch glacier (1957). Sheet 3 portrays an area of pronounced ice moulding of the trough floor, approximating in part to rock drumlins. Part of this area bearing the settlements of Ober and Unter Aletsch lay outside the ice limit of 1850, but most of it has been disclosed by the wastage of the ice tongue during the last century. Individual ribs have lengths of 200–500 m and vertical dimensions of 20–50 m. Since the limits of the ice in a phase of the last glaciation identified as the Daun Stadium have also been mapped, it is possible by drawing cross-sections to scale to see that at this phase the rock drumlin features were covered by 250–500 m of ice. Probably their initiation dates from one of the earlier Würm stadia and these figures are likely to be minima.

From these two estimates we have a fairly clear hint as to the magnitude of the ice thickness responsible for producing streamlined ice-moulded forms – about five times the amplitude of the largest forms in the case of the constructional drumlins south of Lake Ontario, and possibly ten or more times in the case of the erosional features of Aletsch. More evidence is very desirable.

Still more speculative is the answer to the question – by how much has any surface exhibiting conspicuous ice moulding been degraded to produce the forms now seen? One answer to this question was given by W. M. Davis (1900). One of his well-known landscape sketches (see p. 41 of this volume) depicts 'Glaciated knobs in the Central Plateau of France' in an area north of the Cantal; and, noting the absence from this area of the deep soils of nearby unglaciated uplands, he concluded that 'the ice action sufficed to rasp away the greater part of the weathered material and to grind down somewhat the underlying rock, often giving the knobs a rounded profile' (p. 277) – but no more. He reached this estimate well knowing that in the neighbouring valley of the Rhue similar rock knobs are but the much reduced remnants of interlocking valley spurs, and are, in his own words, 'highly significant of what a glacier can do'. Moreover, some of the characteristics of knock-and-lochan topography lead one to wonder whether areas exhibiting this form of modelling had similarly suffered little more than the removal of a preglacial weathered mantle. It would at first sight seem conceivable that selective sub-surface weathering in preglacial times might have etched particularly deeply along just those lines that now form the conspicuous intersecting linear depressions of the present landscape. Yet, though the extent of such preglacial weathering may have had its due influence on the erosive activities of the ice, it seems that the present surface lies everywhere below the former zone of weathering; for nothing but sound rock is found at any altitude and the relative relief of the knocks above the near-by water surfaces is generally more than 200 (and sometimes as much as 400) ft, and the present surface is thus a great deal more accidented than the basal surface of the preglacial zone of weathering is likely to have been.

Possibly we may frame our best estimates of the extent of rock removal involved in the production of ice-moulded features in cases where landforms of simple geometrical character have been substantially overrun. The bounding escarpments of cuestas and plateaux are cases in point.

In western Scotland clear evidence is furnished by the very bold scarp of the Lower Carboniferous lavas that looks down on the Forth lowland near Stirling and its continuations westward. As the Gargunnock and Fintry hills and the Campsie Fells it is a bold cliff-crowned and nearly rectilinear feature that ranks among the more impressive scarps in Britain. It has, in the east, a relative relief of the order of 1200 ft, but this diminishes westward as the base of the lavas rises in altitude. West of Strathblane this reduction in importance permitted the complete overrunning of the Kilpatrick hills by ice from Loch Lomond with the consequent suppression of what scarp feature there had been and the development of knock-and-lochan topography on the upland. In eastern Scotland a rather different case is furnished by the Sidlaws. Here the ice from the west, originating in the Perthshire Highlands, rode up the Sidlaw dip-slope and broke the cuesta into a score or so of very rocky hills. Towards Dundee, and again beyond Forfar, markedly streamlined ridges were developed, and their generally divergent pattern, their independence of strike, and the manner in which both igneous rocks and sandstones are involved, afford convincing grounds for believing that all these forms have been sculptured by the ice from a fairly continuous preglacial upland. In Fife and the Lothians there is much evidence of the same kind but the fame of one type of feature warrants special mention. In the phenomenon of 'crag and tail', which has Edinburgh Castle rock and the ridge of the Royal Mile as its notable exemplar, a small mass of igneous material has been left as an upstanding crag while the more yielding surrounding sediments have been carried away on all sides except in the lee of the crag where they remain as a streamlined tapering tail. Examples of this particular form of ice-moulding are common and it is difficult to see how, even if we assume a measure of positive preglacial relief on the igneous rock, the flanking slopes on the surrounding sediments could, on the average, have been other than fairly symmetrical. The present forms thus imply substantial removal of rock and are true hills of glacial circumdenudation.

A striking instance of a series of fluted forms traversing a single outcrop came to my notice during the war. The Germans were making military use of old underground workings in the rather soft Emscher sandstone (part of the Upper Cretaceous) in the Harz foreland just south of Halberstadt. The sandstone outcrop is repeated by gentle synclinal folding about a west-northwest to east-southeast axis and builds a ridge having the form of a pair of sugar-tongs, closed to the west-northwest and open to the east-southeast. The legs of the tongs are each about 10 km long, they are some 4 or 5 km apart, and the high points of the ridge rise 50–100 m above the general level of the plain which is at 110–120 m. These high points, however, are not found on the crest of a continuous outward-facing escarpment but on a large number of close-set, discontinuous, tadpole-like ridges 10–40 m high and 200–800 m long, with a generally subparallel and east–west disposition (Fig. 8.1). This trend is decidedly oblique to the main outcrops and is even preserved on the north–south portion (the head of the tongs) where the flutings cross the outcrop directly. It is surely to be ascribed to streamlining by ice moving from east to west. The only reflection of structure is seen in the tadpole-like form, for the head of the tadpole is always determined by the scarp side so that in the northern outcrop the tadpoles face east and resemble crag-and-tail forms, while on the southern outcrop they face west and resemble crag-and-tail reversed. Similar features are developed on some other outcrops nearer Halberstadt, though toward the Harz mountains, where ridges of Lower Cretaceous sandstone occur, the dips are so high and the outcrops so narrow that the ridges perforce follow the strike, even if discontinuously, and probably the ice had to do the same. In view of the suggestion already made that such forms in soft rocks may have been covered (like

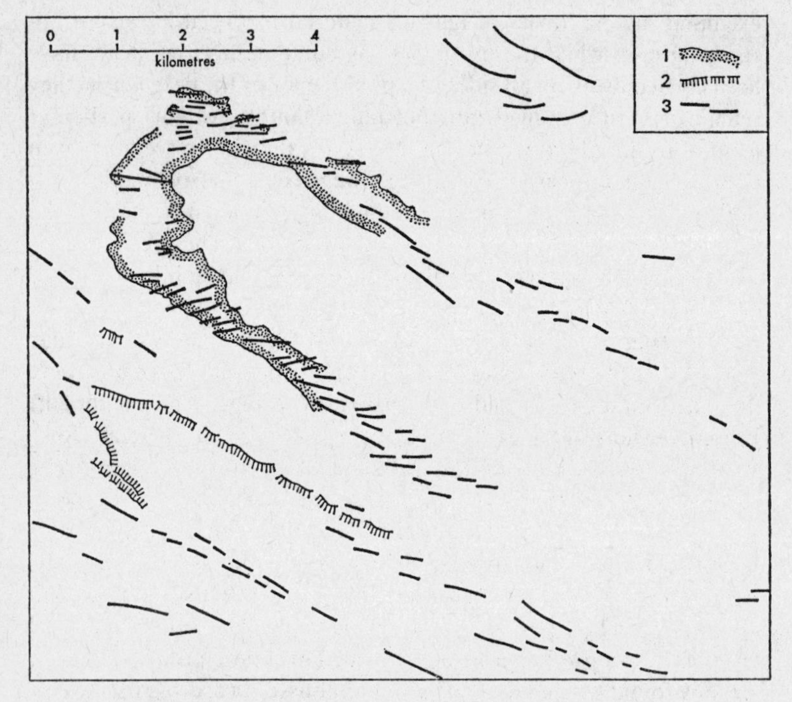

Fig. 8.1 Ice-moulded ridges on Cretaceous and other rocks of the Harz foreland

1. Limits of outcrop (in part) of Emscher sandstone
2. Escarpment of Lower Cretaceous sandstones
3. Ice-moulded ridges

Fairchild's drumlins) by ice having a thickness of the order of five times their own height (say 200 m), it is of interest to note that H. Schroeder (1928) in his description of the geology of the Halberstadt district stated that the 'inland ice' covered the whole area except possibly the sandstone ridges; these, he thought, might have stood up as nunataks.

Perhaps it is permissible here to refer to another case of modification of an escarpment by ice, namely that of the chalk cuesta in East Anglia. Here there is no development whatever of ice-moulded features such as have been engaging our attention, but there is good evidence of profound modification of form and of erosion on the most massive scale. In the Chiltern hills, which are believed never to have been overrun by the ice, the escarpment is a bold feature of very distinctive form, with characteristically a lower bench fronted by the escarpment of the Melbourne Rock, and, set back from the latter, an upper scarp capped by the Chalk Rock that runs forward in a series of bold salients flanked by wing-like spurs. This latter feature, which is repeated beyond Goring gap in the Lambourn and Marlborough Downs, is believed by S. W. Wooldridge and myself (1955) to be an inheritance from the pattern of drainage divides on the late Tertiary peneplain, and, as such, may well have been effaced in districts farther northeast by the transgression of the Calabrian Sea. But the change that takes place in the form of the escarpment as one passes from the Chilterns into the glaciated terrain northeast of the Hitchin gap amounts to much more than this. The scarp of the Melbourne Rock, so conspicuous immediately west of Hitchin, disappears abruptly and entirely, and a feature in this position is not found again until one reaches west Norfolk 50 miles away. Concurrently the main escarpment of the Upper Chalk is replaced by a gentle, partly drift-covered rise to the low boulder-clay plateau of Suffolk. This is set back by many miles from the line of the escarpment farther southwest, and may be traced as a 'height of land' toward Bury St Edmunds, sometimes nearer the southeastern than the northwestern limit of the Chalk outcrop. For upwards of 30 miles in northwest Suffolk and southwest Norfolk even this fails and the Fen streams Lark, Little Ouse and Wissey head in an imperceptible divide some 25 miles east of the Gault outcrop. Around Swaffham the scarp feature reappears and may be followed, paralleled at small distance by the Peddar's Way, to the sea near Hunstanton. This complete failure of a major escarpment over so great a distance cannot be the result of any pre-glacial history since in a normal cycle divides tend to migrate to and to become stabilised along escarpment crests. Rather must it be attributed to massive erosion by ice deploying from the Fen basin. That deployment has been well attested by petrofabric studies of East Anglian tills in recent years, and the massive erosion is witnessed by the vast quantities of chalk incorporated in the boulder clays of Suffolk and Essex.

TROUGHS

Glacial troughs are so numerous and so varied in size and character that it would obviously be convenient if they could be classified in some meaningful way. A possible basis of a classification would be the relation of the trough to the preglacial topography in which it has been developed, and on this basis it is reasonable to recognise four main types, one of which, however, is deserving of further subdivision.

(1) *Alpine type*. The essential feature here is that areas of accumulation are always overlooked by higher ground, which means in practice that they occupy the modified preglacial valley systems. A quite usual arrangement arises from the modification of a convergent series of valley head tributaries to give a convergent series of corries hanging above a main trough. Since this arrangement has been widely regarded as typical of glaciated mountains in general and of the Alps in particular it seems appropriate to call it the 'alpine type'. Probably when the Mittel Aletsch glacier and its tributary glaciers melt away just such a system will be exposed to view. In Britain a good example is afforded by the Ullswater trough and its tributaries from Helvellyn to High Street, and a smaller example by the neighbouring Haweswater system. In western Perthshire the trough containing Loch Voil and overlooked by the Braes of Balquhidder is also a good instance, although some ice entered the basin over cols in the northern watershed. Glen Etive in Argyllshire offers probably our best example, though the ice broke in or out through the bounding watershed at several points.

(2) *Icelandic type*. Here the areas of ice accumulation have been plateau surfaces only exceptionally dominated by higher ground and discharge has been by steep ice falls into the ends of valleys dissecting the plateau. The ice mass of Jostedalsbre in Norway offers an example of such an arrangement still surviving, but in Iceland so great a proportion of the troughs are of this type that it seems proper to speak of it as Icelandic. The almost parallel tributary troughs that lead down northwards from the central plateau to the head of Fnjóskadalur, or the radiating troughs that bite into the plateaux that carried the jökulls of Gláma and Dranga in northwest Iceland, can be cited as type specimens. But it must not be thought that all glaciated troughs in Iceland are like this: excellent examples of the Alpine kind are found in the mountains west and north of Akureyri.

Troughs of Icelandic character are the usual type found in the eastern Highlands of Scotland. We may recognise a series leading down from the Grampian plateau that includes Glen Callater and Glen Muick draining northward to Deeside, and Glen Esk, Glen Clova and Glen Isla draining south or southeast to Strathmore. From the high plateau of the Monadhliath the glens of the Tarff, Brein, Killin and upper Findhorn drain northward and a number of smaller glens southward to Speyside. This type is to be recognised in the Southern Uplands of Scotland, where the valleys of the Scar, Shinnel and Dalwhat tributaries of the Nith offer examples, and most of the valleys that discharged local ice in the Pennines, or in Wales outside Snowdonia, are of this character.

(3) *Composite type*. In both the Alpine and Icelandic types of trough the ice has used and modified a previously existing valley even if the modification has transformed it beyond recognition. But in other cases the ice found no suitably oriented preglacial valleys for its adequate discharge, or not enough of such valleys, and it created new and composite routes, amalgamating portions of earlier valleys with newly created reaches broken through what had previously been divides. It is convenient to classify these composite troughs in categories that reflect the nature of the ice dispersal system of which they formed parts. We may thus recognise troughs of:

(*a*) Simple diffluence, of which there are many Scottish examples, the neatest known to me being Strath Nethy, a distributary branch of the Avon trough in the Cairngorms (Linton, 1949*b*).

(*b*) Multiple diffluence, of which the series of troughs leading away from Loch Fyne through the hills of Cowall (namely, Loch Eck, Loch Striven, Glendaruel and the Kyles of Bute) offer an excellent example.

(*c*) Simple transfluence. This differs from (*a*) in that the new valley reach does not branch sideways from an existing valley across a parallel divide but cuts out of a valley system at its head. The well-known Dalveen Pass is a trough that carried ice from the upper Clyde through the Lowther hills to Nithsdale; Beattock Summit heads a trough that carried Clyde ice into Annandale moving the watershed northward 4 miles in the process; and ice from the Yarrow drainage breached the watershed above St Mary's Loch to spill southward into and to overdeepen the trench now occupied by the Moffat Water. In all three cases passes were opened across the main east–west divide of the Southern Uplands.

(*d*) Multiple transfluence. Ice filling a valley may spill over the divide at its head at more than one point, and additionally elsewhere. Such is the case with Strath Bran in Ross-shire, where the trough of Glen Docharty–Loch Maree leads northwestward, that of Loch Coulin and Glen Torridon westward and that of Strath Carron southwestward from the head of the valley. Additionally ice spilled through the southern watershed by the troughs of Loch Beannacharain, Glen Méinich, Loch Luichart and Loch Garve, into Strath Conon. Multiple transfluence is characteristic of most of the valleys of western Inverness-shire from Loch Morar southward to Ardgour.

(*e*) Radiative dispersal systems. These are associations of troughs that collectively display a marked radiation in plan and formerly functioned as the means of discharge of an area of Highland Ice or a local ice dome. Individually some of the troughs could be classified as Alpine or possibly Icelandic where the ice flowed concordantly to the preglacial valley directions. In other cases the radiate dispersal carried the ice across preglacial watersheds and troughs were cut discordantly through them. Examples of some of these systems are described in Linton (1957) (pp. 130–148 of this volume).

(4) *Intrusive type*. This name is suggested for troughs excavated by ice lobes thrust against the prevailing preglacial gradients from lowlands into uplands. Such ice lobes were 'intrusive' in the sense of deriving from distant sources and entering the local valleys at their preglacial outlet ends. 'Inverse' might be a possible alternative designation.

A clear British example is afforded by Glen Eagles in Perthshire (Linton, 1949*b*). Here ice of west Highland origin, passing eastwards through Strath Allan, pressed heavily upon the northern margin of the Ochil hills. At the phase of maximum glacierisation the hills were substantially overrun, boulder-clay was left in most of the interior valleys, and erratic blocks were stranded high on some of the ridges. But at a late stage, when the upper parts of the hills stood clear, the northern watershed of this small hill mass was pierced by a diffluent ice-lobe at one point only (expectably enough where Strath Allan ice was becoming seriously hemmed in by the steady eastward convergence of the Strath Earn ice upon the margin of the uplands), to create a true glacial trough that is unique in the area. On the displaced watershed at the head of this trough stand the terminal moraines of the intrusive ice-lobe and springing from them a train of outwash gravel descends the valley beyond.

Another notable example of an intrusive trough in Scotland is the buried 'channel' that underlies the lower valley of the Devon for 9 miles along the southern front of the Ochils from Airthrie to Dollar. It has been well explored by numerous borings and in part by a resistivity survey (A. Parthasarathy and F. G. H. Blyth, 1959), and its character and origin have been discussed by Jane M. Soons (1959). Its two deepest basins reach some 350 ft below sea-level, separated by a sill of half that depth, but have no physical connection with the deep hollows below the Lower Carron and the Firth of Forth at Bo'ness. The trough ends rather abruptly about Dollar where the floor rises above sea-level to give way in a couple of miles to a plateau surface above 300 ft with deeply cut postglacial stream gorges and impressive waterfalls. This change occurs, as was noted by Soons, just where the Ochil front recedes to the northeast and divergence of the flowlines of the ice became possible.

SOME CHARACTERISTICS OF GLACIAL TROUGHS

Many features of the transverse and long profiles of glacial troughs are well worthy of discussion, and one of them, the homology of rock steps and *roches moutonnées*, was given extensive discussion by Lewis (1947). Here I wish to take up two points only. The first is simply to remark upon the important morphological consequences of convergence and divergence of the flowlines of the ice. Divergence occurs whenever the ice is released from the constraining effects of the trough walls by an increase in the width of the trough, by a lowering of its sides permitting overspill of the higher ice layers, or by its down-valley termination. Such changes are often manifestly associated with reversed gradients of the trough floor and may be regarded as their usual cause. Conversely convergence is commonly associated with over-deepening. The streaming of large volumes of ice through a constricted passage gives rise to a sort of Venturi effect and the increased velocity has as its result increased erosion, though whether the effects are due solely to abrasion, or to abrasion and plucking together, we have at present no means either of inferring or of observing. Perhaps the effect is most striking where the constriction is itself the result of ice erosion – a glacial breach in a preglacial watershed. Loch Garry, Loch Ericht, Loch Treig and many other Highland lochs all provide excellent examples of this state of affairs.

My second point requires rather more elaboration. It concerns the degree to which the valleyward descent of glacial troughs is concentrated in their uppermost portions, or, put in another way, the strong contrast in gradient that exists between what may be called the headslopes and the main trough. Earlier workers were well aware of this peculiarity. Willard D. Johnson in 1904 spoke of the canyons of the Sierra Nevada of California making descent 'in steps from flat to flat' with 'the tread of the steps in the long stairway . . . greatly lengthened in down-canyon order' and with the initial step (the backwall of the corrie) dominating all, 'the canyons, at their heads' being 'abnormally deep' (see p. 71 of this volume).

This overdeepening is, as Willard Johnson saw clearly, a feature of the mountain canyons. Their heads are steeper and their floors are flatter than those of normal valleys. The little Unteraar glacier of the Alps is only 5 miles long and terminates in the Grimselsee more than 6000 ft above sea-level. The rock wall at its head rises some 2300 ft in little more than half a mile and it is known from seismic soundings that this wall plunges a further 1300 ft at the same steep gradient beneath the ice. The trough then flattens out so that the greatest depth of ice (1457 ft) was found within a mile of the upper limit of the glacier, and the floor has an overall gradient to the Grimselsee (twice locally reversed) of only 1 in 150. There can be no escape from the conclusion that the Unteraar glacier has already deepened its bed in its upper reaches far below what would be possible for any mountain stream. Larger examples are similar if less extreme. The great Norwegian trough of Sognefjord makes the first part of its descent from the peak of Fanaråken at 6000 ft down to the sea at Skjolden in only 13 miles. This fine cascade stairway includes one tread long enough to have a local name (Helgedalen) yet the overall gradient is still 1 in 8. From Skjolden the trough floor descends much more gently with an overall gradient of only 1 in 30 to a depth of 4034 ft in 82 miles, remains nearly flat for 22 miles to reach 4077 ft, and then rises to quite shallow depths (505 ft) just outside the mouth in a further 32 miles. The contrast between head-slope and main trunk could hardly be more striking: less than a tenth of the whole profile is above sea-level yet that small fraction leads up to the watershed of Norway.

The Sognefjord district also provides evidence that glacial troughs have been entrenched far below a topography of essentially preglacial valleys. In the inner part of Sognefjord 80 or 90 miles from the sea the preglacial valley floors lie 2000–2300 ft above sea-level, and their correlatives are visible in the high-level flats at 3700–4000 ft above Fortundalen 140–150 miles from the coast. These figures may not unreasonably be compared with the heights above tide of the valley floor of the Galician river Miño (and its direct upstream continuation the Sil), a river of similar length situated in a rather less boldly elevated but similarly wet environment whose valley was never glaciated – namely, 1000 ft at 85 miles, 1600 ft at 120 miles, and 3300 ft at 150 miles. Doubtless the preglacial Norwegian river had a steeper, and possibly more regular profile than its modern Spanish analogue, but the gradients are clearly of the same order of magnitude. Possibly more immediately convincing are the cases in which the entrenched trough has encroached headward through a preglacial divide into valleys that have suffered much less change because they drained in the opposite direction. At Kinlochleven on the west coast of Scotland the head of the Leven trough has encroached in this way into territory that preglacially drained by the Tummel to some east-coast destination, and though the topography has been modified for some distance east of the original divide the difference of level provides a notable power source for industrial use.

The general conclusion emerges that true overdeepening is not an exceptional but a usual feature of glacial troughs and that far from being characteristic chiefly of the lower ends of well-graded trunk valleys it is found in troughs of all sizes, is most marked in their inland portions where they bite deeply into highlands, and in the cases of great trunk glaciers may achieve magnitudes of 3000–5000 ft. The point I would like to emphasise is the power of a glacier to *bite down* deeply into the land near its sources. This is surely not an effect of mere abrasion but implies mass quarrying of rock far below the local fluviatile base-level and the evacuation of the excavated material down low overall, and often reversed, gradients. This is highly original behaviour and raises large problems as to how quarrying is effected at depths where the rock should be sound and temperature fluctuations minimal. Yet the fact that such behaviour has characterised glaciers of all sizes implies that it is a continuing feature of glacial action and apparently suffers no diminution as the trough grows to large dimensions. This in turn suggests that the growth of the trough is a predisposing factor in the process, and we may perhaps infer that, once a trough has been initiated, for example by pure abrasion under viscous flow, the substitution of a volume of rock by ice of only a third of the original rock density must lead to extensive dilatation of the rock beneath the trough floor, and the parting of the rock along equally extensive sheet joints. The existence of very conspicuous sheet joints, markedly concave upwards, in the naked walls of certain troughs in massive rocks, for example beside Loch Coruisk in the Isle of Skye, or in Tyssedalen in Hardanger, appears to me to be important confirmatory evidence of the existence of such a process.

CORRIES

A good deal of evidence tends to the conclusion that by no means all corries arise from the enlargement of pre-existing valley heads. The numerous glacierets that I have seen clinging to incredibly steep rock walls in Norway and in Antarctica have effectively convinced me that my own earlier notions as to what constituted the minimum dimensions of a 'live' ice body stood in need of revision. The existence of forms intermediate between a torrent gully and something recognisable as a true corrie points in the same direction. Stereoscopic examination of air photographs of the mountains of Andöya in northern Norway reveals both enlarged gullies with no cone of dejection but with the throat of the gully at sea-level, and small and irregular corries partitioned by ribs and benches that are obviously structurally determined. The niche glaciers described by G. E. Groom (1959) as occupying funnel-like hollows on the plateau edges of Bünsow Land in Spitzbergen seem also to belong to the class of transitional or incipient corrie forms. Finally we may note that fully fashioned corries may develop in isolation on suitable hill-slopes even though these are almost unindented by anything deserving the name valley. Such, indeed, is the most westerly of British corries, carved out of the northeast-facing slope of Mount Eagle in 10° 26′ West at the extremity of the Dingle peninsula in County Kerry. Mount Eagle is a residual hill – almost a monadnock – rising steeply on all sides to 1696 ft from a sea-level plain or from the sea itself, and its one corrie holds a lake of nine and a half acres at 780 ft, from which the headwall rises to 1500 ft.

Once a corrie achieves amphitheatrical form it has established conditions that favour its enlargement in plan without change of form. One of Lewis's associates – the geologist M. H. Battey (1960) – has shown that in the wall of a small corrie 400 yd across (Veslgjuvbreen in Jotunheimen) the strike of the major vertical joints passes from east–west to north–south and back again to east–west if the corrie wall is followed from the south side round the head to the north side, and that similar relations characterise a neighbouring corrie (Vesl-Skautbreen). He concludes that these joints arise by spontaneous dilatation of the rock in a direction normal to the corrie wall as the stresses in the rock in that sense are relieved by erosion. Thus the existence of the amphitheatre leads to further amphitheatrical growth by retreat of the head and side walls as long as the masses are removed by the glacier. Battey's observations represent a real contribution to our knowledge of the plucking process, and it is to be hoped that they may be confirmed by similar careful examinations of the joint systems of many corries in a variety of rocks and situations, and extended to a study of the sheet jointing of trough walls and floors where, as I have already suggested, a hypothesis of growth by dilatation jointing appears particularly applicable.

The process of mountain sculpture by corrie growth has presented itself forcefully to the imaginations of many observers since its first recognition by Willard Johnson in 1883. Although the views Johnson later developed on bergschrund-sapping of corrie walls were shown by Lewis in the 1930s not to meet the requirements of the morphological evidence, and are now known also to be physically of limited application, his recognition that corries encroach on the summit upland to produce a remnant whose scalloped outline resembles that of the 'dough on the biscuit board after the biscuit tin has done its work' is in no way invalidated. And in calling the summit upland 'the preglacial upland beyond a doubt' he has the support of later evidence. The deep penetration of chemical weathering along joints revealed in the intersection of upland surface and glacial cliff along the rims of certain Scottish corries, the abundance of core-stones over thousands of acres of the upland, the survival of tors in locations of wide-spaced jointing, the accumulation of quartz-felspar gravels in the hollows, and sometimes the presence of growan with core-stones *in situ* in plateau situations, all point in this direction.

Many workers have commented on the dramatic consequences of the further growth of corries to produce, by mutual interference, sharp-edged divides, pyramidal peaks and cols, and it might be thought that nothing of interest remains to be added. In fact, most of these descriptions owe something to extrapolation, and it is therefore necessary to accept every opportunity to extend our field knowledge of these forms and to keep an open mind about processes. In particular we may look at two suggestions made half a century ago by W. M. Davis (1912) and W. H. Hobbs (1911).

Several text-books have reproduced or adapted a well-known series of diagrams drawn by Davis to illustrate *Die erklärende Beschreibung der Landformen* and it is easy, even in the confines of our own islands, to find examples of the topography depicted by his 'youthful' and 'mature' stages of corrie sculpture. But I venture to suggest that few would undertake to match from nature the diagram labelled 'old age'. It is true that K. M. Ström (1949) has claimed that corrie sculpture may truncate a mountain mass at about the névé line in the manner suggested, but commonly quite a different result follows. This is the progressive wasting of the divides between corries along their lengths, from col to pyramid. The col widens and is commonly overlapped by névé; the pyramid appears to remain unreduced in height but greatly sharpened in form. Numerous Alpine 'horns', with the Matterhorn first and foremost among them, stand head and shoulders above their surroundings to witness to the efficacy of the process of 'arête shortening' and its superiority over continued recession of the corrie headwalls that form the faces of the pyramids. Possibly Battey's observations on the dilatation joints of two Jotunheim corries may afford a clue here. It may be suggested that by the time the corrie walls have intersected to form cols and to outline a residual pyramid, the material of the latter, having been free for some time to expand outwards parallel to three or four free faces, may have reached a state of equilibrium and be subject to only little further change of this kind. The rock core of the pyramid may long remain sound and unjointed. On the other hand, the arête, being traversed in all cases by two sets of intersecting fractures already open, is ripe for disintegration. Certainly I have myself met such an arête. Near Hansen Point on Coronation Island in the South Orkneys my companion, D. J. Searle, and myself had to abandon an attempt to ascend such an arête because it seemed little better than loosely piled angular blocks of the size of pieces of furniture. Later we congratulated ourselves on so doing, for from the beach below we found we could see daylight beneath some of the blocks.

Whatever its mechanism the process of arête shortening has operated to produce some of the most shapely mountains we know. Kyrkja between the heads of Visdalen and Leirdalen in Jotunheimen is well known to many. 'The Pyramid' behind the Hope Bay base of British Antarctic Surveys is another singularly fine example.

The second point that I would like to re-examine is the suggestion of Hobbs (1911) that, as corrie sculpture passes from the stage of his 'grooved upland' to that of his 'fretted upland', the individual corrie 'becomes more irregular in outline . . . not infrequently allowing it to be seen that it is in reality composite or made up of several cirques of a lower order of magnitude'. I believe this to be a mistaken view, and that the development is toward less and not greater irregularity. Hobbs illustrated his suggestion by reference to the tributary glaciers that descend from the Wannenhorn to the Grosser Aletsch glacier, and, thanks to the new maps at 1/10,000 of the Aletsch system it is now possible to examine these little glaciers in a way that would certainly have delighted Hobbs. Each detail of the map suggests that they arose independently and are developing towards amalgamation. The Herbigsgrat, partly covered by névé in Hobbs's day, now stands revealed as a continuous rock wall dividing the Schönbuhl and Wannenhorn glaciers. The subsidiary grats, now not merely overrun but destroyed in their distal portions, clearly separate independent glacierets with their own moraine systems, and in the Schönbuhl glacier these grats articulate with the main ridge at points that are both culminations in elevation and inflexions in plan. Moreover, to assist us in imagining the earlier evolution of these little glacier basins we have three less evolved specimens immediately to the west, now almost free of ice though covered by moraine. They are separated by quite massive rock ridges and have obviously never been anything but independent.

This process of elimination of minor divides is probably quite rapid, for it is found in such mountains as the Cairngorms where corrie sculpture is still youthful. In mountains of mature corrie sculpture such as Jotunheimen and the Alps it is widespread, and various examples may fitly be termed 'shortened grats' (where the distal end has been eliminated), 'interrupted grats' (where beyond an overrun and lowered centre section the ridge rises to a peak at its distal end), 'vestigial grats' (where only the articulation with the main divide remains), and 'subdued grats' or 'smoothed grats' (for those distal portions that have been overrun and ice-moulded before being uncovered). Some interrupted grats lead from the main peak to a small nunatak, or to peaks that, like the three Dreiecks of the Aletsch, were nunataks at a former and higher stand of the ice. Some like the ridges that formerly connected Norway's highest mountain Galdhöpiggen (through Vetlepiggen) to the Galdhöe, or (through other lesser summits) to the Storgrovhöe or to the Dummhöe, terminate in massive remnants of the flanks of the preglacial mountain. Triangular in plan and bounded on two sides by steep walls that are also the trough walls of the separating glaciers, they recall the planèzes of the dissected flanks of the Cantal. Vestigial grats are exceedingly common in the mountains of the Antarctic and are a valuable clue to the interpretation of their morphology. Along many coastlines they survive behind the coastal glacier fringe and were recognised for what they are by Otto Holtedahl in 1928 and their witness is an essential part of his hypothesis of the origin of strandflats. On the inland mountains and nunataks they are ubiquitous and such groups as the Tottan mountains 250 miles east of the Weddell Sea in about latitude 76° S. show little else. Such mountains rise from the inland ice like islands from the sea, but their morphological affinities are not with islands but with monadnocks and the inselbergs of the arid tropics. For unlike islands, which can be and are completely consumed by the waves that outlined them, inselbergs, monadnocks and the residual pyramids of corrie sculpture are not easily reduced by the processes that have created them. If this were not so it would be very difficult to account for the survival of ice-sculptured pyramids in such large numbers in the Antarctic. Today they appear as nunataks though in many cases subdued grats may connect them beneath the ice. Occasionally one finds a single massive round-topped mountain standing up among them and overtopping them by 1000 ft or more, like Mount Taylor in Graham Land, or Mount Bechervaise in Australian Territory, still carrying a remnant of the preglacial surface to testify to its former existence, character and altitude in a scene of wholesale but incomplete destruction. It may well be that in all three types of residual – the monadnock, the inselberg, and the glacial pyramid – destruction remains incomplete for essentially one and the same reason, namely, that the assistance given to weathering and removal in the early stages by dilatation joints is considerably diminished when isolation has actually been achieved.

THE ELIMINATION OF PREGLACIAL VALLEY DIVIDES AND INTERFLUVES

The final theme that I wish to develop concerns the truncation and progressive elimination of the actual divides between preglacial valleys.

In the mountains of the temperate zone, the névé line is at high altitude and the additional volume of ice supplied by the incoming of successive tributary glaciers downvalley tends to be offset, or more than offset, by losses from melting and ablation. There is therefore no general increase in ice volume downvalley and the cross-sectional area of the trough will be no greater in the lower reaches than it was at the névé line. In severer climates where the névé line is close to

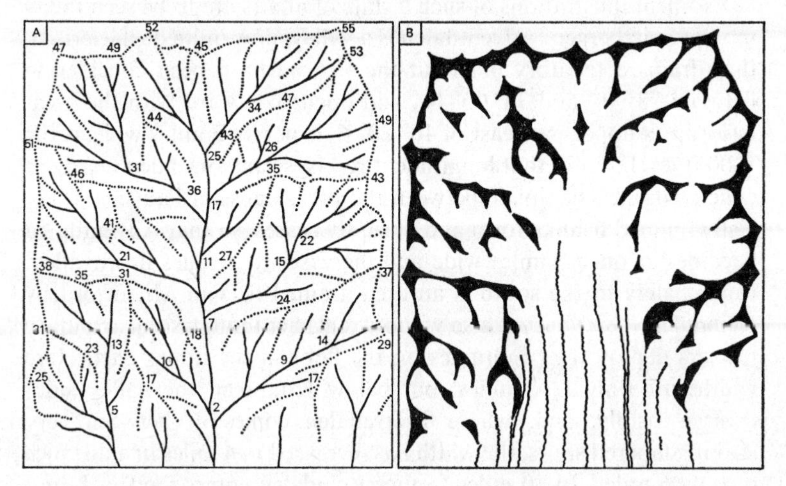

Fig. 8.2 A. *Idealised pattern of streams (full lines) and divides (dotted lines) in an area about 30–40 miles in larger dimension of dendritic drainage on resistant rocks. Figures represent altitudes in hundreds of feet*

B. *The same area after heavy glacierisation with névé line near sea level. Solid black: peaks, arêtes and nunataks. Irregular black lines: moraines*

sea-level this will not be so. The incoming of tributary glaciers will not be offset by wastage and the total ice volume will increase and require an increasing cross-section. The two halves of Fig. 8.2 attempt to visualise the effects of such a state of affairs upon the landscape. Fig. 8.2A depicts a mountainous tract of country with a dendritic pattern of preglacial drainage and dichotomously branching divides. It is supposed to cover an area about 25–35 miles, and, being developed in hard rocks and strongly elevated, stream gradients are high. On glacierisation every valley engenders its own glacier and as the névé line descends to lower and lower levels these all become confluent, and since the tributary ice-streams retain their identity great increase in width ensues. This, it will be seen, involves the cutting back of many of the preglacial lateral divides by several miles, with, it is suggested, scouring on the upstream sides giving markedly rounded or stream-lined forms to their terminations. And, whereas the preglacial spurs on the two sides of the main valley as it approached the coast were only 5 or 6 miles apart, the separation of the last visible divide remnants in Fig. 8.2B is of the order of 15 or 20 miles.

Excellent illustrations of such a state of affairs are to be seen today in west Spitzbergen. Adventdalen is a valley, now quite deglacierised, that drains a territory of about the size we have been considering between 78° 10′ and 78° 15′ N., and opens westward with the great Isfjord. Its upper part east of 16° 20′ E. is nearly 2 miles wide at the 1000 ft level, but in that longitude it receives Janssonsdalen nearly as wide as itself. The divide between the two – Janssonshaugen – takes a streamlined fishlike form and abruptly ceases, so that Adventdalen becomes at once 3 miles wide and increases to 4 miles downvalley. Immediately to the south is an area drained to Van Mijenfjord by Reindalen. This valley is also without a glacier though some tributary glaciers deploy their moraines on the floor in the upper part. Here widths are only 1½–2 miles, but below Kokbreen the valley sides diverge visibly, and, where Tverradalen comes in between very striking tapered spurs, the width has increased to 4 miles in a distance of only 6 miles. In 10 miles farther Reindalen enters the fjord on a front 7 miles wide, and just at its entrance receives Litledalen from the east from behind a long spur spectacularly rounded, tapered and recurved away from the main valley. These cases are both taken from the region of level-bedded Tertiary sediments with their strikingly tabular relief. But Dunderdalen developed on the resistant, massive Hecla Hook formation shows the same characteristics even more forcefully. In less than 10 miles it widens from only 1 to more than 3 miles by the steady recession of its streamlined walls and the tapering away to nothing of the 1400 ft high Slettfjellet in 2½ miles. A much more real contrast is in fact to be found between all the valleys that we have been considering (which are all without glaciers) and the area of greater precipitation a little farther south between 77° 0′ and 77° 20′ N. (where the landscape is still heavily glacierised). Torellbreen exemplifies strikingly the conditions suggested in Fig. 8.2B. As the eastern tributaries of Vestre Torellbreen come in – Profilbreen and Skoddebreen – they suffer a sharp deflection (through 60 or 70 degrees) downvalley, and their moraines are forced into parallelism with the larger glacier, while the spur ends (Klockmann-fjellet, Fortanna and Krakken) are conspicuously rounded and offset. A little to the southeast, and rising in the same névés, are the glaciers Hansbreen, Paierlbreen and Mühlbackerbreen separated by the Sophiekammen and Luciakammen. Both glacier troughs and ridges are strongly influenced by the north-northwest to south-southeast strike of the Hecla Hook rocks, but rock structure has little to do with the most striking feature of both ridges. Both have formerly had greater breadth and between their lateral spurs held low-level corries. One or two of these still survive, but most have been reduced to trifling dimensions by the paring away of the lateral spurs by the trough glaciers. The truncated spur ends have been driven back till they have coalesced and the ridges as a whole have assumed stream-lined form. Luciakammen has become fishlike in plan with its nose at Luciapynten: Sophiekammen is essentially similar, but still possesses a long subdued grat leading to Fannytoppen.

If Luciakammen and Sophiekammen in west Spitzbergen must be regarded as divides between parallel valleys that have been shorn almost completely of their lateral spurs and had even their valley head corries truncated, there is no reason to think that they do not have counterparts elsewhere, and even counterparts that exhibit the phenomena in question to an even greater degree. In the Mawson hinterland in Australian Antarctic Territory (between 62° and 63° E., and 67° and 68° S.) the Mount Henderson group, and the Masson, David and Casey ranges are all that remains above the ice of a system of major divides. They run almost due north and south, the Masson and David ranges for some 40 miles and the Casey range for 30 miles, with 10–20 miles of northward streaming ice between each pair. This ice has obviously overrun or carried away not merely spurs but all the lateral divides and the ranges are reduced virtually to lines of single pyramids rising 800–1500 ft above the ice prolonged by strings of nunataks farther south.

In Graham Land, where there is no inland ice to stream outward in broad ice rivers 10–30 miles across that carry all before them, there is nevertheless ample evidence of the paring away of divides between parallel glaciers. But because the glaciers are relatively small and head against an interior plateau the resulting geomorphological forms are strikingly individual. They may be introduced by consideration of a small example in the neighbourhood of Hope Bay. Between the troughs of a small unnamed glacier and the much larger

Depot glacier lies The Steeple. This rock remnant between two troughs of different size, depth and curvature, clearly once extended much farther as a ridge separating the two glaciers. The former continuations of The Steeple are lost without trace – pared entirely away by Depot glacier and its smaller neighbour. But before it disappears entirely it takes the form of a low ice-moulded ridge – a subdued divide again – barely 100 ft above the ice. From this point the distal end of The Steeple towers up like the prow of a ship. This is the edge along which the slightly convergent walls of the two troughs intersect, and as glacial abrasion continues this edge must retreat upvalley.

Here is a principle of prime importance for the geomorphology of much of Graham Land and the associated island groups. Numerous minor divides have been shortened to a condition like that of The Steeple; in a few cases Steeple-like buttresses can be seen projecting through the present ice surface with a glacially subdued rock spur in front of them. Such is the buttress between the West and East Stenhouse glaciers in Admiralty Bay in the South Shetlands. In many more cases the shortening has advanced so far that only a vestigial divide remains, as may be seen in Ezcurra Inlet, also in Admiralty Bay. We thus have shortened and vestigial divides comparable to our shortened and vestigial grats but resulting in the main from different processes. But the most striking illustrations of divide-shortening are to be found along the margins of the lofty (3000–6000 ft) interior plateau of Graham Land. This plateau, like Willard Johnson's surface in the High Sierra of California, is the preglacial surface scalloped by the inbiting of concave glacier headwalls. But the destruction of the divides between neighbouring troughs has apparently been faster than the recession of the headwalls of the troughs into the plateau. On the south-eastern and less heavily glacierised side, divide remnants as much as 5 or 6 miles long may remain. They may carry narrow slivers of plateau at their proximal ends and be scalloped by second-generation corries that have been truncated and left hanging by the glacier flowing past them. In their middle portions they are gracefully sculptured with horn peaks and cols, while their distal portions are lower, with only isolated pyramids. Beyond – nothing! Yet occasional nunataks far out amid the ice surely testify to the former extent of the preglacial landscape in this direction, and excite the imagination to wonder how much has been removed. On the north-western and more heavily glacierised margin of the plateau that destruction has gone even further. For dozens of miles along the Palmer Coast it is clear that only vestigial divides remain. Far out in the ice toward the Weddell Sea are some of the nunataks to which they were once joined.

CONCLUSION

The forms of glacial erosion are to be regarded as the products and the witnesses of an association of processes whose nature and mode of operation we understand imperfectly enough, but whose efficacy is such that even in the short span of time represented by an ice age they can remove mountains. These processes are in all cases the work of ice in motion, in some cases moving downward and outward from a corrie wall, in others streaming through troughs that may be enlarged till they are 5000 ft deep like Sognefjord, or scores of miles wide like some Antarctic troughs, or in yet other cases the moving ice is deploying widely over marginal lowlands to reach the sea. I am myself so impressed by some of these results that I wonder whether moving ice is not in some areas the most efficient agent of bulk erosion and removal in limited time that is operative in the wearing down of the lands. One recalls John Playfair's assertion that 'For the moving of large masses of rock, the most powerful engines without doubt which nature employs are the glaciers'. Yet away from the great avenues of movement, or the slopes where the descent of even small masses is rapid, the ice lies sluggish and essentially passive. Even in the Antarctic, ice may be seen to be both powerfully erosive and effectively protective. Work done in fact is the expression of energy expended – a conclusion that any physicist would have predicted.

REFERENCES

Eidg. Landestopographie, Aletschgletscher, 1/10,000, Blatt 3 (Wabern-Bern, 1960).

Battey, M. H. (1960) 'Geological factors in the development of Veslgjuvbotn and Vesl-Skautbotn', in W. V. Lewis (ed.), *Investigations on Norwegian Cirque Glaciers*, Royal Geographical Society Research Series No. 4, 11–24.

Dana, J. P., cited in G. K. Gilbert, op. cit., 207.

Davis, W. M. (1900) 'Glacial erosion in France, Switzerland and Norway', *Proc. Boston Soc. Nat. Hist.*, xxix 273–322.

Davis, W. M. (1912) *Die erklärende Beschreibung der Landformen* (Leipzig) Figs. 151–5.

FAIRCHILD, H. L. (1907) 'Drumlins of central western New York', *New York State Museum Bulletin*, CXI 391–443.

GILBERT, G. K. (1910) *Harriman Alaska Series*, vol. III: *Glaciers and Glaciation*.

GROOM, GILLIAN E. (1959) 'Niche glaciers in Bünsow Land, Vestspitzbergen', *Journal of Glaciology*, III 368–76.

HOBBS, W. H. (1911) *Characteristics of Existing Glaciers* (New York).

HOLTEDAHL, O. (1929) 'On the geology and physiography of some Antarctic and sub-Antarctic islands', *Scientific Results, Norwegian Antarctic Expedition*, III 146–68.

JOHNSON, WILLARD D. (1904) 'The profile of maturity in alpine glacial erosion', *Journal of Geology*, XII 569–78.

LEWIS, W. V. (1947) 'Valley steps and glacial valley erosion', *Transactions and Papers, Institute of British Geographers*, XIV 19–44.

LINTON, D. L. (1949*b*) 'Watershed breaching by ice in Scotland', *Transactions and Papers, Institute of British Geographers*, XVII 1–16.

LINTON, D. L. (1957) 'Radiating valleys in glaciated lands', *Tijdschrift van het Koninklijke Nederlandsche Aardrijkskundig Genootschap*, LXXIV 297–312.

LINTON, D. L. (1962) 'Glacial erosion on soft-rock outcrops in central Scotland', *Biuletyn Peryglacjalny*, XI 247–57.

NYE, J. F. (1952) 'A comparison of the theoretical and measured long profile of the Unteraar glacier', *Journal of Glaciology*, II 103–7, Fig. 1 (based on P. L. Mercanton and A. Renaud, in works cited by Nye).

PARTHASARATHY, A., and BLYTH, F. G. H. (1959) 'The superficial deposits of the buried valley of the river Devon near Alva, Clackmannan, Scotland', *Proceedings of the Geologists' Association*, LXX 33–50.

RICHTER, E. (1896) 'Geomorphologische Beobachtungen aus Norwegen', *Sitzungsberichte der Wiener Akademie Math.-Naturw.*, CV 152–64.

SCHROEDER, H. (1928) *Erläuterungen zur geologischen Karte von Preussen, Blatt Halberstadt, 2307*.

SOONS, JANE M. (1957) 'The geomorphology of the Ochil hills', unpublished Ph.D. thesis, University of Glasgow, 114–17.

STRÖM, K. M. (1949) 'The geomorphology of Norway', *Geographical Journal*, CXII 19–27.

WOOLDRIDGE, S. W., and LINTON, D. L. (1955) *Structure, Surface and Drainage in South-east England* (London).

9 Glacial Erosion in the Finger Lakes Region, New York State

K. M. CLAYTON

IN areas that have been heavily glaciated the ice may locally have created an entirely new landscape; it may have erased the preglacial land surface and created a new pattern of hills and valleys in its place. Such new relief, consisting of glacial troughs separated by smooth, streamlined interfluves, can well be mistaken for a landscape of sub-aerial origin, modified by the passage of the ice-sheet, but still bearing the pattern of valleys and water partings that existed before. Some glaciated landscapes are indeed of this kind, but there are other areas where careful examination of the evidence suggests a complete discordance between the preglacial valley pattern and that cut by the ice. D. L. Linton (1957) has suggested that the radial arrangement of the troughs of the Western Highlands of Scotland or of the English Lake District was created by the ice, and not inherited from preglacial times. The ice moved transversely to the line of the earlier river valleys, and did so with enough power to cut deep troughs that transect the dismembered remnants of the preglacial valleys. Only the most careful reconstruction can establish the nature of the preglacial relief, and the degree of transformation it has undergone. The Finger Lakes area of upstate New York would seem to offer a landscape of this type. In this paper an attempt is made to establish that the whole relief near the Finger Lakes is the work of ice, and is quite independent of the form of the preglacial landscape. The destruction of that earlier landscape is almost complete, and no attempt is made at this stage to try and reconstruct it in any detail.

The Finger Lakes lie in the north-western part of New York State, U.S.A., south of Lake Ontario, and along the northern edge of the Allegheny Plateau. Their long, narrow and very deep basins have been recognised to be the result of glacial erosion for well over 50 years, as by R. S. Tarr (1909). Indeed, since they occur in a part of the United States that is devoid of any alpine landforms, they have become a classic case of glacial erosion. It is an exaggeration to suggest, as O. D. von Engeln (1961) has recently done, that these lakes are unique (remarkably similar features are found in the Traverse City area of Michigan, and Lake Memphremagog on the

Vermont–Quebec border, to name but two examples), but it is certainly true that the erosion is very localised. It is by no means confined to the valleys alone; thus N. M. Fenneman (1938) writes: 'Obviously these lakes are in the valleys of preglacial streams, and just as obviously their basins above the water level have been modified by moving ice.' More recently, R. F. Flint (1957) has expressed much the same view: 'These valleys were first cut by preglacial streams flowing north. They were widened, greatly deepened and straightened by subglacial erosion when the Laurentide Ice Sheet repeatedly filled and overran them from the north.' In fact, there is no direct evidence that these valleys occupy the sites of preglacial streams flowing northward in this embayment in the Allegheny escarpment. It is something that has been inferred from the pattern of the valleys, which converge towards the north, and from the shape of the embayment in the escarpment which has the form that might be expected had it been cut back by these hypothetical obsequent streams.

During the course of a two-year stay in upstate New York, the writer had the opportunity to examine the landforms of the glaciated Allegheny Plateau. From field work and map analysis over an area of some 10,000 square miles in New York and northern Pennsylvania, some idea of the form and intensity of glacial erosion was built up. It is apparent that there was relatively little glacial erosion in the south, and that this was fairly localised. To the north, both the amount and the location of glacial modification undergoes a progressive change, until in the zone immediately south of the Finger Lakes almost all the landscape except a few fragments of plateau has been profoundly altered by the passage of the ice. Every valley has been turned into a glacial trough, and they cut through the upland in every direction, looking like the intersecting passages of some unroofed ants' nest. It is thought that this increase in the intensity of erosion includes the Finger Lakes themselves, and that here a new valley system has been cut across the earlier landscape; the lakes seem to be incised transversely across a south-westerly directed drainage pattern. They were initiated as the southward-moving ice overrode the escarpment, forming what D. L. Linton (1963) has called 'intrusive troughs'. Support for the morphological evidence is found in the nature of the glacial tills along the New York–Pennsylvania boundary. The Binghamton drift, once thought to mark an independent ice advance (P. MacClintock and E. T. Apfel, 1944), marks the dominant movement within the ice-sheet, and consequently the direction of trough erosion.

THE GLACIATION OF THE ALLEGHENY PLATEAU

The Allegheny Plateau (Fig. 9.1) descends by a series of rather poorly developed escarpments to the drumlin-covered plain south of Lake Ontario. The plateau itself is far from uniform in height; although

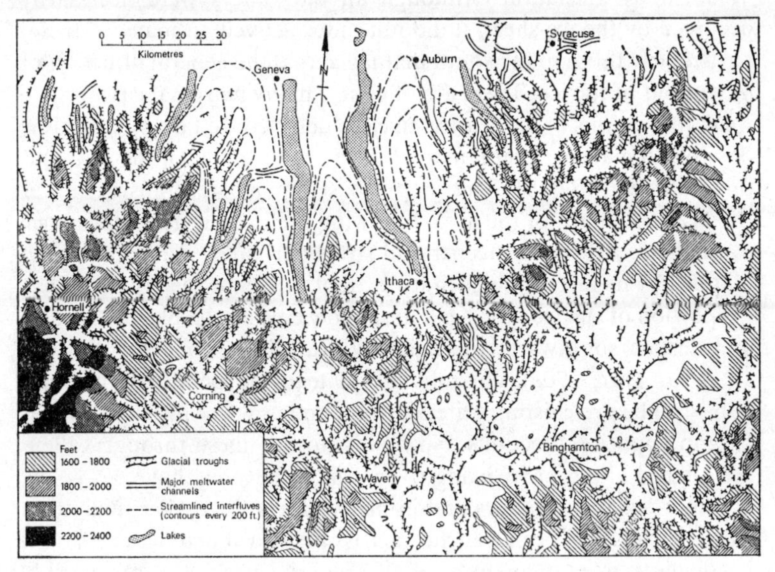

Fig. 9.1 The Finger Lakes and the adjoining part of the glaciated Allegheny Plateau

appreciable areas southwest of Corning rise above 2200 ft, an equally extensive area around the River Susquehanna lies between 1400 and 1600 ft. On the whole, the surfaces (or surface) that form this plateau are quite independent of the structure of the area; the general plateau level is related to the main trend of the present drainage. Thus there is a zone of higher summits along the line of the water parting between the streams flowing south to the Susquehanna and those flowing north, through the Finger Lakes, to Lake Ontario; a second line of summit levels rather higher than the general plateau surface lies south of the east–west section of the Susquehanna and its western tributary, the Cohocton–Chemung. It must be realised that these levels refer to the glaciated plateau surface, but as has already been suggested, away from the main glacial troughs the plateau seems little affected by glacial erosion, a view supported by C. D. Holmes (1937, 1952) and more recently by E. H. Muller (1963). Where the glacial troughs are not too closely spaced, the broad plateau form must be very similar to that overrun by the ice.

By no means all the valleys of the area have been appreciably modified by glaciation. Although all the valleys were necessarily occupied by the ice-sheet, it did not move actively in many of them, particularly those in the south that lay across the general direction of movement of the ice. There are of course many marginal cases where weak indications of ice erosion may be detected, but despite this it is possible to make a general map of the glacial troughs of the area and this is done in Fig. 9.1. This map is based on a careful inspection of the 1:24,000 maps of the area, following preliminary reconnaissance in the field. All of the steeper and straighter valley walls have been mapped by means of the hachure symbol. It will be seen that, while a proportion of the troughs head in shallow cols on the plateau surface and deepen southward, the majority are 'through valleys' (W. M. Davis in Tarr, 1905), cut into the plateau independently of what appear to be pre-existing water partings and now linking streams that flow in opposite directions. No doubt most of these through-valleys are aligned along pre-existing river valleys, but they have cut below the floors of those valleys and in many places the earlier, high-level water parting has been trenched across by glacial action. The result, particularly in a zone south of the Finger Lakes, is a remarkably complex network of through-valleys. The origin of this valley pattern was described by Holmes in 1937, and there seems no reason to differ from his interpretations. In particular, there seems every reason to agree with his thesis that the ice moved along all these valleys in a complex pattern of flow, rather than the alternative hypothesis that the area was invaded by ice from successive directions, each glaciation picking out the valleys that led in the general direction of the ice movement.

Of the several objections to this second hypothesis, two are particularly significant. First, there is considerable evidence, from striae and other evidence of the detailed direction of the ice movement, that the ice did in fact change direction at some of the valley junctions (Holmes, 1937; also MacClintock and Apfel, 1944). Secondly, it should be noted that erosion from several directions might be expected to produce one series of valleys superimposed upon another series trending in a different direction. In fact, places where two valleys cross in this way, each continuing in its former direction, are very uncommon; and the crossings are usually offset, as at Van Ehen and Spencer, south of Ithaca (Fig. 9.2). For the ice to have followed such an angular course implies a facility to turn sharp corners, which seems to be most consistent with the view that ice flowed through the whole of the valley complex at the same time.

South of this zone where most, if not all, the valleys were occupied by moving ice, is a rather different landscape. It is most clearly seen south of the Chemung–Susquehanna line. Here glacially eroded valleys are far less common; they are in fact restricted to a few that cross the east–west water-parting that lies near the State line: for example, the valley from Apalachin Creek over to the North Branch of the Wyalusing Creek, or from the valley above Nichols (west of Owego) south to the Wysox Creek (Fig. 9.2). These troughs are of the through type described by Holmes (1937) – leading up a north-flowing valley and breaking across into a south-flowing valley. In each case the amount of erosion has not been very great; the col is still at quite a high level, and although the valley sides have truncated spurs, they are far less steep-sided than the troughs farther north. Alongside these rather scattered troughs are many stream valleys showing no apparent modification by ice-action. With the possible exception of the Susquehanna valley itself, no valleys lying across the general direction of movement of the ice have been affected; indeed, by no means all of the valleys lying parallel to the ice-movement show any clear signs of erosion by the ice.

To the north of the Susquehanna (where it flows from east to west) there is a gradual transition to the well-developed through-valleys already described. Between the north–south through-valleys occur troughs that head on the plateau in shallow cols and deepen southward towards the Susquehanna. Good examples are the Nanticoke and Genegantslet Creeks. Within a few miles north of the Susquehanna some of the cross (roughly the east–west) valleys are also eroded by ice. Then, south of the Finger Lakes, is the zone, usually some 10 miles in width, with a fully developed through-valley complex, the plateau having been reduced to many relatively small, isolated fragments. This is a sequence which from south to north shows zones of increasingly intensive glacial erosion. In all cases the

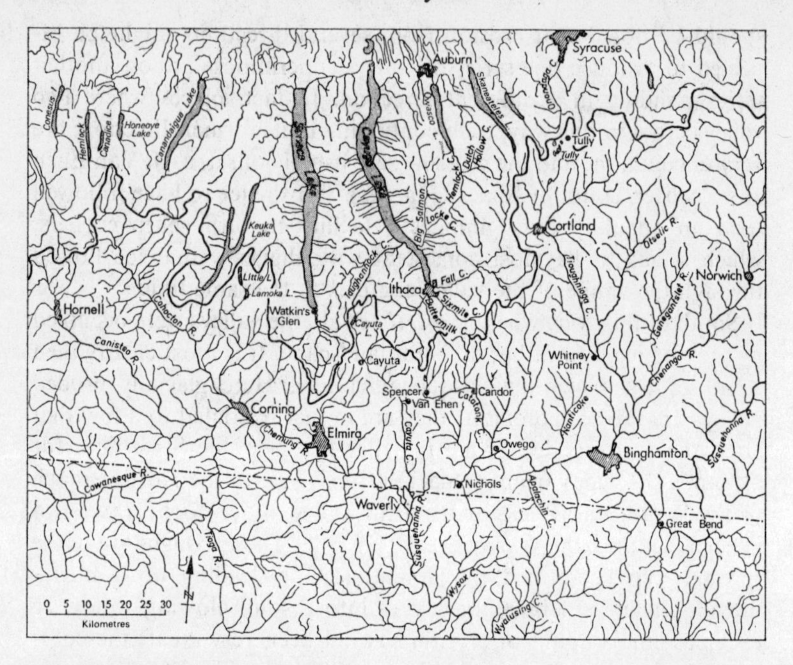

Fig. 9.2 The Finger Lakes: place names mentioned in the text. The solid line indicates the water parting between the Great Lakes and the Susquehanna drainage. The dot–dash line is the New York–Pennsylvania state boundary

erosion is concentrated in the valleys; the plateau may be increasingly segmented, but there seems no reason, even in the complex through-valley zone, to regard it as seriously lowered by the passage of the ice. There, even more clearly than farther south, the plateau consists of relatively level surfaces terminated by the steep trough walls. Where occasional beds of rather more resistant rocks dip gently southward, the plateau is diversified by small cuesta-like features presenting steep faces towards the north. They were surely adjusted to structure by preglacial erosion and offer supporting evidence of the absence of glacial modification of the plateau. Had the ice been actively eroding here, we might have expected to find these features smoothed on the north, with their steeper, plucked sides facing south. Everywhere, then, the glacial erosion was concentrated in the valleys, but it became increasingly selective southward as it became less severe. From moving along almost all the valleys in the north, the ice tended to flow only along those that led in the direction in which it was moving; south of the Susquehanna indeed, it only moved rapidly along those that gave easy passage across the higher part of the plateau surface.

THE FINGER LAKES

The Finger Lakes lie, on the whole, north of the main zone of the through-valleys. To some extent the smaller lakes to the west and the east lie in or on the northern margin of the zone, but in general the zone is free of trough lakes. To a certain extent this is the result of the infilling by glacial outwash of any lake basins that may have existed south of the Valley Heads moraine. The long stand of the ice at or near this line was accompanied by a very large volume of outwash that poured southward down the valleys and eventually down the Susquehanna. Thus although well records in valley sites are few and far between, there are several that show an infilling of the order of several hundred feet; one of the deepest records of valley fill comes from Norwich in the Chenango valley where a well in the centre of the valley passed through 418 ft of drift before reaching bedrock. These thicknesses are exceeded in the case of the Finger Lakes valleys; even Honeoye, one of the smallest, has a valley fill of 646 ft at the north end of the lake (from unpublished well data received by N. Y. State Gas and Oil office), but it is clearly not possible to rule out the occurrence of closed rock basins south of the Valley Heads moraine. Certainly the Valley Heads moraine is often the site of the present water parting and MacClintock and Apfel (1944) termed these 'valley stopper' moraines. Since we have no adequate subsurface data for most of these valleys, nothing is known of the relationship between the valley floor and the form of the bedrock profile. However, the investigation by F. Durham (1958) of the Valley Heads moraine, near Tully (south of Syracuse), suggests that we need not regard each valley segment of the moraine, with its southward-sloping outwash train, as necessarily obscuring a reverse (northward) slope in the bedrock floor. It will later be suggested that this is in fact rarely the case.

There is a remarkable symmetry in the form of the Finger Lakes. With the exception of the westernmost lakes they lie in a radiating pattern, the two largest lakes in the centre of the group. The troughs occupied by these largest lakes (Cayuga and Seneca) are known to be extremely deep. In each case the lake bottom descends below sea level, while boreholes have shown that the fill of unconsolidated sediments occupies a far deeper trough in the solid rock, the floor of which has not been reached. (The greatest recorded depth was a well put down in the middle of the delta at the south end of Lake Seneca and abandoned, still in drift, at a depth of 1080 ft – H. S. Williams, 1909.) Although the lakes show the line of the glacial trough perfectly well, it is difficult to map the troughs by plotting the steep valley walls as was done farther south. The immediate sides of the lake are not particularly steep (except locally towards the southern ends) and there is no clear break at the edge of the plateau such as

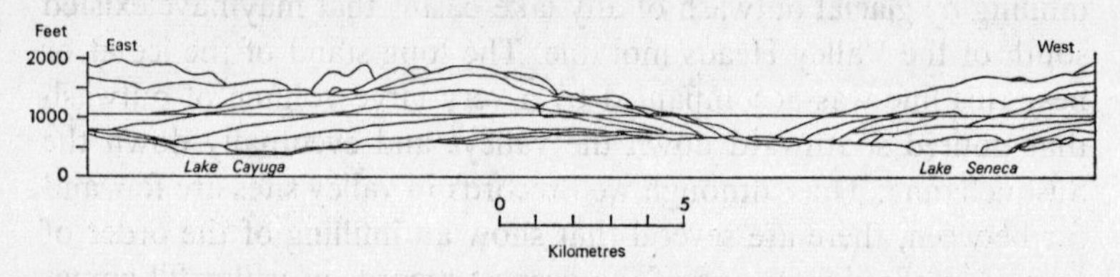

Fig. 9.3 Projected profiles of the central Finger Lakes interfluve, looking from the north

was noted farther south (Fig. 9.1). Indeed, about midway along the lengths of Lakes Cayuga and Seneca the land rises smoothly from one lake to a rounded crest and falls equally smoothly to the other lake, the whole interfluve forming a broad convexity that is increasingly steep towards the lakes (Fig. 9.3). The most striking change, very readily seen when the maps of the lake area are compared with those of the zone to the south, is the way in which the intersecting through-valleys die out along the lakes. In the Finger Lakes area the cross-valleys are much less frequent and less well marked than in the southern zone, and most of the lake-free troughs that do occur (as between Lake Cayuga and Lake Owasco) lie parallel to the lakes.

We have seen that the usual explanation of this simple pattern of sub-parallel valleys has been to suggest that the lower elevation of the area led to a strong flow of ice through this breach in the northern escarpment of the Allegheny Plateau. The previous erosion of the scarp is attributed to the headward erosion of a series of north-flowing obsequent streams, the same streams providing the valleys which so concentrated the glacial flow that the deep Finger Lakes troughs were cut. The hypothesis carries with it certain unsolved problems. Why should such a series of troughs be succeeded to the south by the complex zone of through-valleys? Would it not be the tendency of the ice-sheet, having reached the higher plateau, to follow even more strongly those routes that led southward and continue the line of the Finger Lakes? Why should the ice divide and meander through this intricate trough pattern, only to return farther south to a preference for those routes that led it most easily southward? To these questions V. E. Monnett (1924) added the problem that difficulties lie in the way of accepting an earlier northward-flowing drainage, even were we to suppose that a set of vigorous obsequent streams would have such remarkably straight and converging valleys. He also regarded the suggested alternative of a southward-flowing former drainage as equally improbable: his own solution, at least in the case of the lakes east of Cayuga, was to suppose that they were cut along the line of opposed right and left bank tributaries of a trunk stream that drained westward towards Lake Cayuga. His explanation has the merit that it brings the occasional cross-valleys into play, e.g. the line of Locke Creek, Hemlock Creek, Dutch Hollow (Fig. 9.2); these tend to appear in other reconstructions of the preglacial drainage of this area as remarkably long pseudo-subsequent tributaries of the obsequent streams.

An alternative hypothesis of the origin of the Finger Lakes would seem to follow from the northward increase in the intensity of glacial erosion that we have already noted. The most intensely eroded zone that has been described was the through-valley section, with the ice moving through the plateau on a complex course, dividing and reuniting, but everywhere moving along the valleys rather than across the plateau remnants, where it was thinner and moved relatively slowly. Any increase in the severity of the ice erosion would lead to the elimination of these separate barriers to free ice movement, and a general overrunning of the whole area by active ice, plateau and valley alike. Fenneman (1938) draws attention to the smooth, eroded valley slopes, but the possibility that this might imply a remodelling of the whole landscape rather than just the smoothing of former stream-channelled slopes was apparently not considered. 'Ice erosion is suggested by the forms of the exposed slopes. They are wonderfully smoothed, increasing in steepness towards the lakes with a gently convex curvature. Except for a few postglacial ravines, the contour lines are almost as straight as the lakes. For many miles together the absence of tributaries is almost complete. Not only were the lateral ravines of preglacial time filled, but the divides between them were smoothed off. Nothing but a glacier could have made the sides of these lake basins so simple.' This quotation can be matched by a description of the ice-moulded slopes in Strath Allan, Scotland. 'The upper slopes are astonishingly regular. The contour of 1000 ft runs uninflected for over 8 km, and higher contours up to that of 1500 ft run parallel to it. Down these slopes numerous small streams descend at right angles to the contours, flowing south or south-southeast, and without the slightest incision' (Linton, 1962). These two sentences could well be a description of the eastern valley side of Lake Cayuga, although there the 1000-ft contour runs uninflected for over 25 km, and the minor postglacial streams descend *westward* to the lake. Here too we must agree with Linton that these forms indicate 'considerable ice erosion'. That the ice in this area moved actively across the whole landscape is not, then, in doubt. The point at issue is whether we are to restrict our interpretation of the erosion to a general smoothing and simplification of a former valley pattern, or whether the ice cut a new surface largely independent of the preglacial surface.

It appears that the extension of erosion to the plateau allowed a simplification in the direction of the ice movement as the plateau became of less importance. No longer need the ice meander through a complex of valleys as in the through-valley zone; here the whole ice-sheet could move in the same general direction. In this way the erosion of those valleys lying transverse to this direction would decrease, and the lowering of the plateau would gradually lead to their elimination. Those valley segments that run in the same direction as the ice flow would already have been linked up into continuous troughs by the destruction of the interfluves, since this is a feature of the through-valley complex; they would then be further lowered, no doubt at a faster rate than the intervening plateau. The result would be a stream-lined and lowered upland surface with shallow, poorly developed, cross-troughs to the south where erosion has progressed less far, and few or no cross-troughs to the north. We might expect too that under these conditions the rapidly moving ice would tend to fan out as it moved south, since it would move more powerfully than the ice on either side. This would tend to impose a radial pattern of troughs (and in this case lakes) which is clearly analogous to the radial movement of more isolated ice-domes described by Linton (1957). A further feature common to this case and the area described by Linton is the way in which the ice flowed outward remarkably independently of the shape of the surface over which it moved. It is this imposition of a trough pattern almost regardless of the pre-existing landscape, breaking through divides and so linking formerly separate drainage basins, that makes the identification of the preglacial elements so difficult. In particular, it is suggested that because the ice streamed

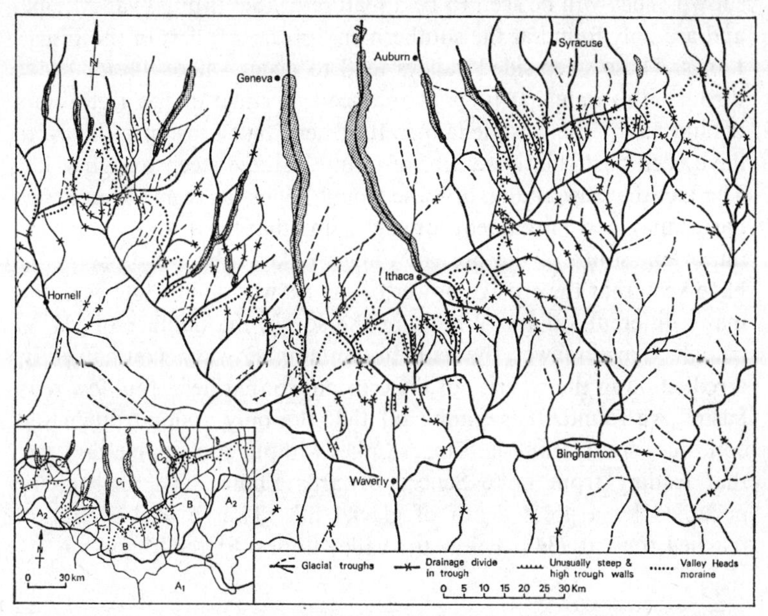

Fig. 9.4 The glacial troughs of the Finger Lakes area. 'Unusually steep and high trough walls' are defined as those that maintain a slope of more than 1 in 3, through a height range of more than 500 ft. Inset: Regional subdivisions according to the intensity of glacial erosion (for key to letters see text)

through here it consequently lowered this part of the escarpment, rather than that the ice moved through this section because the escarpment was already lower than it is to the east or the west.

The present landforms of the central part of the Finger Lakes show the features described in the previous paragraph very clearly. The streamlined form of the interfluve is shown on Fig. 9.3, where a series of projected profiles is drawn for the ridge between Lakes Seneca and Cayuga. From these profiles, too, can be seen the higher plateau, broken by cross-valleys, which lies to the south. The map of the glacial troughs (Fig. 9.4) shows the absence of cross-troughs and the remarkable sub-parallel arrangement of all the main troughs. The erosion of the plateau has had the effect of reducing the break of slope at the top of the trough walls and, in addition, of reducing the angle of the trough walls. These are steepest where the valley erosion, compared with the erosion of the plateau, is most intense, and this can readily be seen in Fig. 9.4, where these steep trough walls are shown. They will be seen to be a feature of the through-valley zone, and are only found at the southern end of the smallest of the Finger Lakes. These steep-sided valleys tend to occur where the ice broke through the water partings at the head of some of the right-bank tributaries of the Susquehanna. It is here that erosion is likely to have been particularly severe, resulting in these steep trough walls. The fact that few, if any, of these trough segments seem to have rock basins may be an argument for suggesting that these were never very important south of the present Finger Lakes. The reason seems to have been that the erosive power of the ice was divided between too many different valleys. Where, in the south, the number of glacial troughs is again fewer, the reduction in number was accompanied by a reduction in the severity of the erosion, so that here, too, few rock basins are found. It is significant that the only major trough that continues south from the Finger Lakes without appreciable division, that leading from Lake Seneca towards Elmira, is known to be occupied by a great depth of glacial fill. (Thus a well at Elmira reached rock at 734 ft below the valley floor – Fairchild, 1925.)

THE ORIGIN OF THE SUSQUEHANNA WATER PARTING

This recognition of a series of concentric zones of increasingly severe glacial action may provide a useful subdivision of the extensive glaciated Allegheny Plateau described by Fenneman (Fig. 9.4 inset). There are three main zones, the southern zone (A 1) marked by occasional through-valleys, with a subsidiary zone found occasionally to the north (A 2), where troughs head in the open plateau and lead south to the Susquehanna. North of this is the through-valley zone (B), beyond that the streamlining and trenching of the severely eroded Finger Lakes (C 1). Again there is an intermediate zone here where some of the smaller Finger Lakes reach into the through-valley zone (C 2). The approximate boundaries of these regions are shown on the inset to Fig. 9.4. A distinctive feature is the very sudden termination of deep scouring at the southern end of many of the

Finger Lakes. As has already been noted, sub-surface data are poor, but there is frequently a marked change in trough form which would seem to indicate a rapid rise in the bedrock floor. Only in one place can this be demonstrated in any detail – this is in the (lake-free) upper Onondaga valley west of Tully. The conclusion of Forrest Durham (partly based on seismic data) that the bedrock floor south of the Valley Heads slopes southward towards Cortland has already been noted. Recent bores north of the moraine have demonstrated that the valley floor (at about 600 ft above sea-level) is underlain by at least 518 ft of recent fill (unpublished well data, N.Y. State Gas and Oil office); Durham (1958) estimated the bedrock floor at Tully Lake to be about 975 ft above sea-level, giving a net rise of the rock floor of at least 900 ft in 4 miles.

For the greater part of its length, the Valley Heads moraine is the present site of the water parting between the Lake Ontario and the Susquehanna drainage. Previous workers have been reluctant to regard a divide along the line of a Wisconsin moraine as of more than recent origin, and it was this that led MacClintock and Apfel (1944) to suggest that there might be deep deposits of rift obscuring former through-valleys. Since one of the most striking of these cases, that at Tully, has been shown to be a rock barrier with a relative relief of 900–1000 ft, and since the general hypothesis adopted here suggests a rapid diminution of the depth of trenching along this zone south of the Finger Lakes, it seems time to abandon this view. The present divide remains, however, totally unrelated to the preglacial divide at the head of the Susquehanna drainage, for the rock divide is the result of deep scour on the northern side during the glacial period, and is almost as young a landscape feature as the Wisconsinan Valley Head moraine itself.

THE BINGHAMTON DRIFT

South of the Valley Heads moraine, two distinct types of drift occur. In northeast Pennsylvania and on much of the plateau north of the State line in New York, the drift is referred to as Olean. It is rather weathered, drab-looking drift, with few far-travelled constituents. Locally, particularly in the main valleys north of Binghamton, there are drift deposits which are brighter in appearance when seen in section, with abundant igneous and limestone erratics. The status of this fresh-looking Binghamton Drift has been discussed by several authors, and while at times it has been regarded as a lithological variant of much the same age as the Olean, evidence has also been put forward that it is a younger advance, although clearly pre-dating the Valley Heads drift. MacClintock and Apfel (1944) put the evidence for a separate Binghamton advance most strongly, and drew an ice front that lay south of the valley deposits of Binghamton type, and which they regarded as the limit of an advance younger than the Olean and older than the Valley Heads. The evidence was almost entirely the contrast in the lithology of the two drift sheets.

More recently J. H. Moss and D. F. Ritter (1962) have rejected this conclusion, deciding that the Binghamton drift is but a local variant of the Olean drift sheet, and that there is no evidence that it was deposited by a later ice advance. They reach this conclusion on the basis of a wider range of evidence than MacClintock and Apfel, examining topographic expression, pebble lithology, heavy mineral content, drift texture and pebble orientation counts. As they stress in their account, and as is clearly shown in their distribution maps, the Binghamton drift is a deposit of the main valleys and does not occur on the plateau. Nor is it found in all the valleys; it is best developed in those valleys that head in the limestone belt to the north – the 'through' valleys. This is in contrast to the findings of MacClintock and Apfel (1944), who stated that 'If there are any lithologic variations related to either through-valleys or transverse valleys influencing the ice flow they have not been found.' It may be that their conclusion reflects the difficulty of finding exposures on the plateaus between these valleys; certainly the Olean-type drift here is very poorly exposed, and nearly always very thin. Moss and Ritter find the Binghamton drift to have a 'tongue-like distribution', extending down the main valleys from the area of the Valley Heads moraine. While a distribution of this type might be explained in terms of a more lobate ice front than that envisaged by MacClintock and Apfel, Moss and Ritter reject such a solution, and instead favour two other factors. These are (1) the presence in these valleys of river deposits rich in limestone pebbles, which have been incorporated into the glacial drifts there, and (2) 'a differential rate of flow between valley and upland' ice.

A necessary corollary of the account of the glacial erosion in this area that has been advanced in this paper is that the ice movement was concentrated along the valleys of the main area of the Allegheny Plateau. The width of these more rapidly moving zones became greater to the north, as did the zone of effective erosion, until in the area of the Finger Lakes, the whole landscape was covered by moving and eroding ice. Even so, here too movement was more rapid, and erosion more severe, along the present lakes. Such differential movement is supported, not only by the form of the relief, which was the reason for advancing the concept, but also by the distribution of the Binghamton drift. This is the deposit of this rapidly moving valley ice. The plateau areas were also covered by the ice, but it moved far more slowly, was incapable of much erosion, and left behind a drift that is always thin, and always composed predominantly of local rocks. Northward, these areas of relatively stagnant ice became more and more limited, until in the area of the Finger Lakes the entire area was converted to a streamlined surface capable of carrying the huge discharge of ice from the Lake Ontario basin.

REFERENCES

DURHAM, F. (1958) 'Location of the Valley Heads Moraine near Tully Center determined by preglacial divide', *Bull. Geol. Soc. America*, LXIX 1319–21.

FAIRCHILD, H. L. (1925) 'The Susquehanna River in New York and evolution of western New York drainage', *Bull. N.Y. State Museum*, CCLVI 36.

FENNEMAN, N. M. (1938) *Physiography of Eastern United States* (New York: McGraw-Hill).

FLINT, R. F. (1957) *Glacial and Pleistocene Geology* (New York: Wiley).

HOLMES, C. D. (1937) 'Glacial erosion in a dissected plateau', *American Jour. Sci.*, XXXIII 217–32.

—— (1952) 'Drift dispersion in west-central New York', *Bull. Geol. Soc. America*, LXIII 993–1010.

LINTON, D. L. (1957) 'Radiating valleys in glaciated lands', *Tijds. van het Kon. Aard. Genoot.*, LXXIV 297–312.

—— (1962) 'Glacial erosion on soft-rock outcrops in Central Scotland', *Biuletyn Peryglacjalny*, XI 247–57.

—— (1963) 'The forms of glacial erosion', *Trans. Inst. Brit. Geogr.*, XXXIII 1–28.

MACCLINTOCK, P., and APFEL, E. T. (1944) 'Correlation of drifts of the Salamanca re-entrant', *Bull. Geol. Soc. America*, LV 114–16.

MONNETT, V. E. (1924) 'The Finger Lakes of Central New York', *American Jour. Sci.*, 5th ser., VIII 33–5.

MOSS, J. H., and RITTER, D. F. (1962) 'New evidence regarding the Binghamton sub-stage in the region between the Finger Lakes and the Catskills', *American Jour. Sci.*, CCLX (2) 81–106.

MULLER, E. H. (1963) 'Reduction of preglacial erosion surfaces as a measure of the effectiveness of glacial erosion', *Proc. I.N.Q.U.A. Conference, Warsaw*.

TARR, R. S. (1905) 'Drainage features of Central New York', *Bull. Geol. Soc. America*, XVI 229–42.

—— (1909): see WILLIAMS, H. S., *et al.* (1909).

VON ENGELN, O. D. (1961) *The Finger Lakes Region* (Ithaca, N.Y.: Cornell U.P.).

WILLIAMS, H. S., *et al.* (1909) 'Description of the Watkins Glen – Catatonk District, N.Y.', *U.S. Geologic Atlas*, folio 169, 33 pp.

10 Measurements of Side-slip at Austerdalsbreen, 1959

J. W. GLEN and W. V. LEWIS

ONE of the most interesting problems in the theory of glacier flow concerns the extent to which glaciers slip past the bedrock. The problem is of theoretical interest because, until recently, there was no theory which enabled the amount of this slip to be predicted, and the recent theory of J. Weertman (1957) has still to be tested experimentally. It is also of interest in any consideration of the mechanism of glacial erosion, for clearly a glacier which is stuck fast to its bed will be a less active eroding agent than one which is slipping and carrying with it debris which can scour the bedrock. However, the slip of a glacier over its bed at depth is a difficult quantity to measure, and it is therefore of interest to see how fast glaciers slip relative to the rock in those places where the interface is easily accessible, that is, where the glacier side is sufficiently clear of snow and moraine for the ice–rock contact to be visible from the surface.

J. D. Forbes (1842), with his usual discernment, noted two features of glacier movement that have appeared in our measurements on Austerdalsbreen. He observed that on the Glacier des Bossons, near Chamonix, the difference of velocity between the centre and the sides was least at the higher part and most at the lower part of the glacier. He further noted (1843) that the Mer de Glace moved unevenly, quicker in warm, fine weather than in cold weather. Erratic movements on glaciers have been recorded again and again, for example by H. B. Washburn and R. P. Goldthwait (1937) on the South Crillon Glacier, by W. R. B. Battle (1952) in Greenland, and by M. F. Meier (1960) on the Saskatchewan Glacier. Abrupt changes both of direction and speed of movement have been measured accurately on Veslgjuv-breen, a cirque glacier in Jotunheimen (J. E. Jackson and E. Thomas 1960), with firn moving down from its *névé* fields in two directions almost at right angles; but in the small, simple, neighbouring cirque glacier Vesl-Skautbreen, ice within 1 m of the bed moved uniformly at about 1 cm per day (J. G. McCall, 1960).

In 1956 one of us made some measurements of side-slip at two places immediately below the ice falls which join to form Austerdalsbreen, a distributary from Jostedalsbreen, Norway (Glen, 1958).

It was then found that at these sites the flow was of the same order of magnitude as the velocity at the centre of the glacier. Since the foot of an ice fall is a place where, according to Nye's theory of glacier flow (1952), large amounts of slip on the bed are to be anticipated, and also are regions where *Blockschollen* movements of the kind discussed by W. Pillewizer (1950) are most marked, these results were not unexpected.

Since 1956, members of the Cambridge Austerdalsbre Expeditions have made various measurements of the glacier, including, more recently, measurements on the main part of the glacier tongue and on the snout. In the course of these it was thought desirable to look for sites farther down-glacier than those in the earlier side-slip measurements, and to test whether at these stations also there was appreciable slipping past the side walls. In his measurements on Austerdalsbreen in 1937, Pillewizer had deduced the variation of velocity across the glacier and had found that, although the velocity was roughly constant over the middle parts, it did appear to fall off rapidly as the edge was approached. A similar result was obtained from stake measurements made by the Cambridge Austerdalsbre Expedition in 1958; it was thus of interest to see whether in this region the slip of the glacier past its bed had fallen to zero, or whether a small amount of definite slip remained.

EXPERIMENTAL DETAILS

Three sites were found on the left bank of the glacier (the east side) and one on the right bank (the west side) at which the ice–rock interface was relatively unencumbered by moraine. The location of these sites is shown in Fig. 10.1. Their locations will also be described here in some detail, as they help to account for some of the variations in the amount of side-slip from place to place. On the left bank the uppermost site was immediately downstream from a fan of avalanche material which fell to the glacier from a hanging valley. The avalanche snow and ice persists through the year and so forms a small but definite part of the accumulation of the glacier. Just downstream from the fan which had formed in the previous winter we found a place where the ice flowed past rock with only a few encumbering boulders on its surface. Upstream the fan provided a region of relatively thicker ice, while downstream there was a region of comparatively gentle surface gradient. This site was called 'East Fan' station.

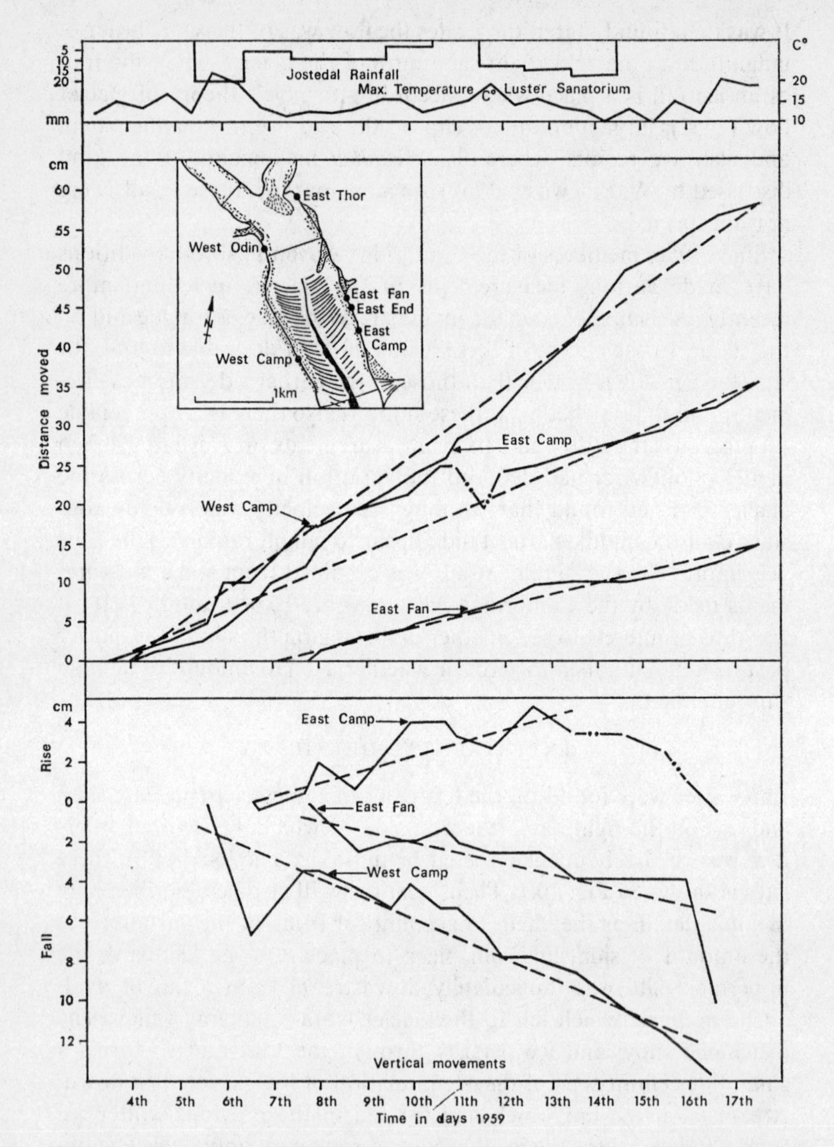

Fig. 10.1 Measurements of horizontal and vertical movements of ice past rock using the first method, with graphs of rainfall and temperature and, inset, map of Austerdalsbreen showing location of sites used for measuring side-slip

The second site was a few hundred metres down-glacier from the first, where a transverse line of stakes used for velocity measurements reached the east margin of the glacier; it was therefore called 'East End' station. There was slightly more debris on the ice here, and several large boulders had entered the gap between the ice and the rock. The surface slope of the glacier was still relatively slight, though the ample covering of rock debris had protected the margins of the glacier from ablation so that the surface lay above the general level of the debris-free ice a hundred or more metres from the side wall.

The third site on this side of the glacier was a few hundred metres still farther downstream, and, being approximately opposite the camp site used by the expedition on the medial moraine, was called 'East Camp' station. It was just up-glacier from a point where the valley widened and the surface gradient increased suddenly. The site may be on the downstream edge of a rock bar. There was still a considerable amount of debris on the glacier surface, but sites for drilling into relatively clean ice could still be found.

The new site on the right bank of the glacier was downstream from the transverse line of stakes, and was approximately opposite the position of the camp site and so was named 'West Camp' station. This margin of the glacier was considerably more covered with moraine than at the sites on the left bank, and the surface stood higher above the main debris-free glacier. This site was not ideal, but was the only one available on that side of the glacier. The stake position was at the foot of a steep slope of ice. Rocks slipped down this slope from time to time as their supporting ice melted away. This hazard made it difficult to preserve stakes for any length of time, and a stake was in fact destroyed by such a rock slide, curtailing measurements at this station.

In addition to these new sites, measurements were also made at 'West Odin', a position close to that adopted in 1956 on the right bank of the glacier. The actual site, L, used in 1956 could not be relocated, but the new site was within a few tens of metres of the old position. The object of this station was to see whether velocities of flow in this region were still of the same order of magnitude as they had been three years previously.

The methods of measurement adopted were the same at all the new sites, and two independent series of measurements were made at each, each of us being responsible for one method. In the first method (W.V.L.), use was made of the gap between glacier and side wall that was present at every site. A hole about 2 m long was bored through the ice into this gap and a 3 m stake of square section with 2·5 cm sides was inserted through it. The end of this stake was pointed, and the movement of this end was measured relative to pencilled marks on the rock wall. However, in order to avoid damage to the stake from stones falling down the gap between ice and rock, and to reduce ablation effects, some stakes were pulled back into the ice between measurements. The end of the bore hole near the rock was in the shade and the other end was covered each day with snow and rock debris after the stake had been pulled in. This reduced melting in the hole to a minimum so the stake remained firm enough not to wobble, and yet could be pushed forward to touch the rock for a measurement to be made. One night when the stake had not been withdrawn, a boulder fell down the gap and snapped the end off. Despite precautions, slumped debris filled the gap in wet weather and had to be removed to reveal the stake for measurement. The main advantage of this method of measurement is that it records both horizontal and vertical components of motion very close to the wall; the disadvantages are that melting tends to make the stake fit only loosely into its hole, causing uncertainties as to its correct positioning, and also that a rotation of the ice near the wall may be misinterpreted as a translational motion.

The second method of measurement (J.W.G.) was that described in the earlier paper (Glen, 1958); a stake was inserted vertically into the ice a convenient distance away from the edge and its distance from two paint marks on the rock was measured. In this way the movement of the top of the stake relative to the rock wall can be calculated. The advantages of this method are that a vertical stake suffers less from ablation than one in a sloping hole projecting right through the ice, and that measurements of velocity components perpendicular to the wall can be made. The disadvantages are that the point of measurement is necessarily some definite distance away from the actual margin, and that vertical components of motion, or more accurately components parallel to the wall and perpendicular to the line joining the paint marks, cannot be determined.

EXPERIMENTAL RESULTS

At all sites definite slip past the sides was recorded by one of the two methods, and at only one site, 'East End', was it not recorded by both methods. Here the first method of measurement gave neither horizontal nor vertical movement during the 7 days from 6–13 July. As this stake was sheltered from falling rock it was left in position throughout this time, so the lack of movement cannot be attributed to errors in re-positioning. However, melting revealed a large boulder, mainly enclosed within the ice but also extending across the gap to the rock wall, just down-glacier from the stake. This may account for the null readings in one of two ways; either it may have been holding

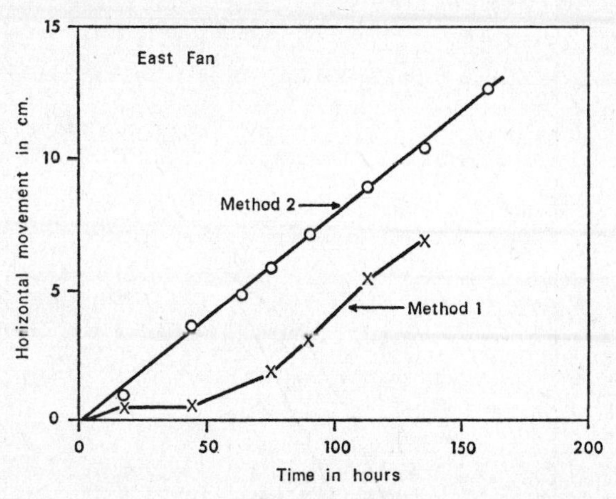

Fig. 10.2 Graph of movement of ice at 'East Fan' station as measured by the two methods

back the ice at this point, or else the stake, which touched the boulder at one point, may itself have been pivoted back relative to the ice as the ice pushed its other end past the rock, so as to leave stationary the point of the stake. The first alternative seems to be the more likely. At all the other new positions results were obtained by both methods and it is thus possible to compare them. Figs. 10.2–10.4 show the total movement measured by the second method compared with the horizontal component of movement parallel to the wall obtained by the first method for the time during which both methods were employed. It will be seen that the first method appears to give more erratic movements than does the second method. The variations are probably too large to be solely due to errors in the first method such as the melting of the hole containing the stake, and so represent in part genuine fluctuations in velocity at the very edge of the ice. These fluctuations are smoothed out at the distance of about 2 m, which is the approximate distance of the stakes from the ice edge at all the sites for the second method. These measurements thus seem to show fairly definitely that, at about 2 m from the edge of the ice, the flow is very uniform, and is not much influenced by weather conditions, while closer to the ice edge more erratic movements are taking place which may be related to weather conditions.

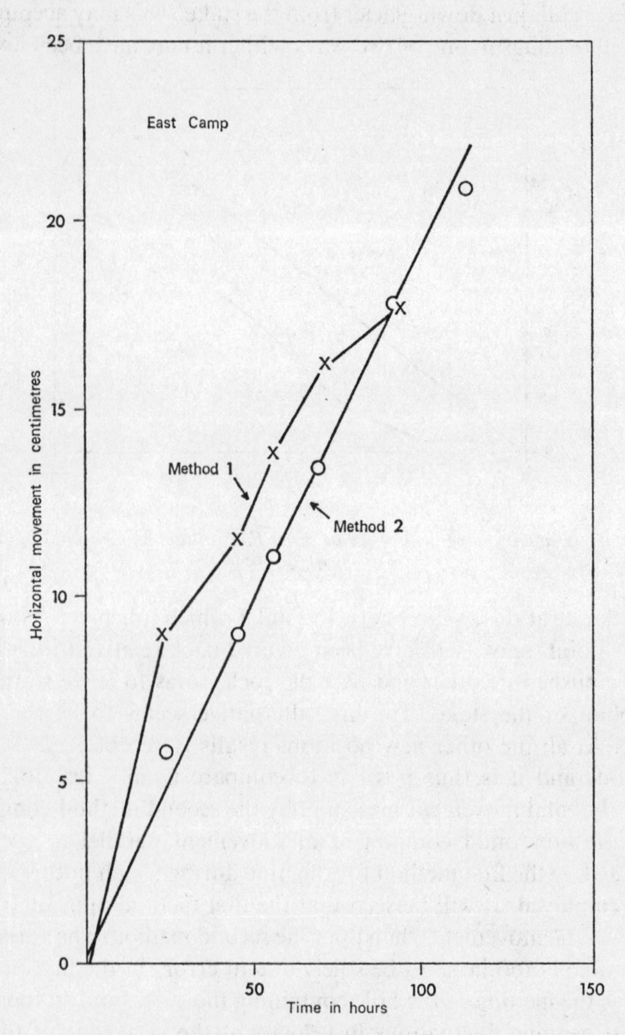

Fig. 10.3 Graph of movement of ice at 'East Camp' station as measured by the two methods

Fig. 10.1 shows the measurements made by the first method over the whole period of observation, a longer period than that during which both measurements were being made. Minor suggestions of halts can be seen on several of the traces. The reverse movement shown in the 'West Camp' record on 11 July may partly be due to an inaccurate measurement at this station when the observer was alone at the site

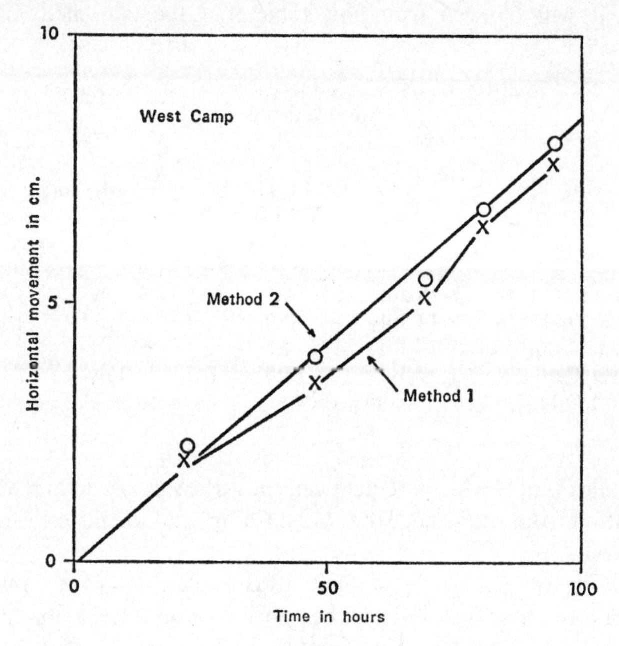

Fig. 10.4 Graph of movement of ice at 'West Camp' station as measured by the two methods

in the rain. The system of measurement at this station, the first that was chosen for work by this direct method of observation, was rather crude. A plumb-bob had to be suspended with the point touching the point of the stake and held by hand in mid-air; then this hand holding the plumb line string had to be moved horizontally in a fixed direction until it touched the rock wall. Then a pencil mark was made along the string. The fixed direction was taken from the corner of a boulder some metres out on the ice, and in the absence of a second observer the determination of this direction was only approximate. A partial correction was carried out when the station was occupied by two observers because the second observer could readily see if the pencil line was out of the vertical as given by the plumb line when viewed from the corner of the boulder.

The average rate of movement varied from site to site. The results obtained for the average rate of horizontal flow using the first method and for the total flow rate (ignoring the unmeasured component in the vertical direction) using the second method, are shown in Table 10.1. It will be seen from this table that the two methods give

Table 10.1 Measurements of average rate of horizontal side-slip on Austerdalsbreen

Site	Method 1		Method 2	
	Period	Velocity cm/day	Period	Velocity cm/day
'East Fan'	7–18 July	1·5	12–19 July	1·9
'East End'	6–13 July	0·0	12–19 July	1·6
'East Camp'	3–17 July	4·2	13–18 July	4·3
'West Camp'	3–17 July	2·5	12–16 July	2·0
'West Odin'	Not recorded		15–18 July	19·6

velocities which, while not identical, show the same general pattern of variation from site to site; thus 'East Camp' had the fastest slip of all the new sites.

The movement at 'West Odin' was somewhat slower than the 30 cm/day found at L in 1956, but since a second determination that year gave 25·6 cm/day, there seem to be large variations in this area and it cannot be said with certainty that the velocity at a given point has decreased over the years. It is clear, however, that the velocities at the lower sites are considerably smaller. They are to be compared with velocities of about 13 cm/day in the centre of the glacier in the neighbourhood – measured at the transverse line of stakes. The slip past the side is thus some 10–20 per cent of the velocity at the centre of the glacier, a smaller percentage than that obtained just below the ice fall (Glen, 1958).

Vertical movements by the first method are also recorded in Fig. 10.1. These, like the horizontal movements with this method, were usually accurate to ± 3 mm judged from repeating measurements during one observation. The movements are again erratic, and this appears more markedly than with the horizontal measurements because the net vertical movements are smaller. 'East Fan' showed a fairly consistent fall averaging 3·5 cm in 10 days with relatively small departures from this average. 'East Camp' station, on the other hand, rose at an average of 6·5 cm in 10 days and showed large variations. The suspected rock bar beneath this station may account for part of this upward movement, but a rise is natural in the lower reaches of glaciers where ablation is great, provided that the mass balance of the glacier is in approximate equilibrium over a period of years. The surface gradient is that which is sufficient to enable the glacier to carry downstream its normal load of ice. Ablation during the hot summer months lowers the surface, and thins the glacier, which would reduce its velocity and still more its discharge unless the glacier thickens due to compression until equilibrium between supply and discharge is approximately restored.

The departures from average rate of movement on these two traces are closely in phase, if allowance is made for the departures being consistently smaller at 'East Fan' station than at 'East Camp'. Several of these erratic movements start later at 'East Fan' station although it is farther up-glacier. Perhaps the smaller disturbing forces at this locality take a little longer to make themselves felt.

'West Camp' station fell at an average rate of 7 cm in 10 days, i.e. slightly quicker than 'East Camp' station on the opposite side of the glacier rose. There is little or no sign of a rock bar on this side of the glacier and the considerable height of this station above the general level of the glacier may have tended to increase the downhill movements as melting progressed. The movements were very erratic during 5, 6, 7 July. Thereafter they became much steadier as did the horizontal movements at this station.

DISCUSSION OF RESULTS

The most significant feature of the results obtained is that slipping of the glacier past its side wall does occur in this region of Austerdalsbreen at all sites at which measurements were taken, and that the amount of slip varies markedly from site to site. In particular the slip at the middle station on the left bank was less than that both above and below it. The rate of slip is probably determined by the nature of glacier flow in the vicinity of the site. At 'East End' station the glacier was relatively level and of constant width; however, downstream from 'East Camp' station the glacier passes out of a constriction and then widens and loses height relatively rapidly. At such a position the ice can be imagined to be squeezed past the rock, giving an enhanced rate of side-slipping. This helps us to understand the erosive power of glaciers at such places.

The more variable rate of flow observed by the first method using a stake projecting through the ice may, as has been mentioned above, be partly due to experimental errors. If, as seems likely, it indicates a genuine variation in the rate of flow very close to the edge of the glacier, this may be due to irregularities in the side wall, to boulders incorporated in the ice and jammed against the rock near the stakes, and to sudden melting due to heavy rain and high temperatures. Table 10.2 shows precipitation and temperature conditions at the nearest meteorological stations during the period of observations. Fanaråken, at an altitude of 2062 m, is 50 km to the east; Luster Sanatorium, at an altitude of 484 m, is 25 km to the east, and Jostedal, at 370 m, is 16 km to the east. Temperature figures are not available for Jostedal. Rainfall data for Jostedal, and maximum temperatures for Luster Sanatorium are shown in Fig. 10.1, as these stations are near to Austerdalsbreen and at roughly the same altitude. They confirm an experience gained in the field, that heavy rain, and to a lesser extent high temperatures, influence the glacier movement. The sudden forward lunge of 'West Camp' station by 5 cm between 14.00 hours and 22.00 hours occurred during a very heavy rainstorm on 5 July. The temperature was also high, reaching the maximum during the period of observation of 22·5°C at Luster Sanatorium. This was by far the quickest and the greatest movement measured at any of the side-slip stations. It was accompanied by the greatest vertical movement which, however, continued when the horizontal movement had ceased. This brief pause was followed by another large horizontal movement on 6 and 7 July, making a total movement of more than 12 cm in 48 hours. The ice then settled down to move a little more than a third of that speed for the next 10 days. The vertical movement was also excessive and erratic during and immediately after that period of heavy rain and high temperature, falling abruptly nearly 7 cm and then rising 3 cm rather suddenly.

The 'East Camp' station, though much steadier in its movements, also surged forward at this time, and it accelerated again during the period of fairly heavy rain on 7 and 8 July. 'East Fan' station, where measurements began on 6 July, also lurched forward and downward on the evening of 7 July when the 'West Camp' station paused for a

Table 10.2 Temperature and precipitation data for the Jostedal–West Jotunheim area

By courtesy of the Director, Det Norske Meteorologiske Institutt

Date	Temperature				Rainfall			Snow depth
July	Fanaråken		Luster Sanatorium		Fanaråken	Jostedal	Luster Sanatorium	Fanaråken
	Min. °C	Max. °C	Min. °C	Max. °C	mm	mm	mm	cm
1	1·3	7·0	10·2	18·2	0·0	0·9	0·5	85
2	0·2	4·0	9·0	13·8	7·1	0·0	0·7	75
3	0·5	6·0	9·6	14·9	2·0	1·4	0·3	70
4	1·0	8·0	9·5	17·2	1·1	0·0	2·4	65
5	3·0	9·0	10·2	22·5	1·4	0·8	0·4	60
6	1·5	8·0	12·0	19·7	22·0	20·3	17·8	55
7	0·0	3·0	9·2	13·0	8·0	4·5	4·5	50
8	0·6	4·6	9·0	14·3	2·5	7·9	3·6	50
9	2·0	4·6	12·4	16·7	0·7	8·0	2·0	45
10	−1·5	2·4	8·7	15·0	3·8	1·3	0·0	40
11	−1·0	5·0	7·9	20·5	0·0	0·0	0·0	35
12	−0·2	4·5	10·3	16·5	21·5	10·7	10·8	30
13	−1·2	3·4	8·6	14·5	14·6	12·5	5·6	25
14	−1·0	1·0	8·0	13·7	8·0	15·8	11·1	25
15	−5·0	6·0	4·8	15·5	0·0	0·0	0·0	25
16	1·0	5·5	8·2	21·1	0·2	0·0	0·0	20
17	0·0	5·0	12·0	19·0	0·0	0·0	0·0	20
18	1·4	6·0	12·0	21·2	0·0	1·4	3·1	15

second time after its second forward lunge. It is instructive to note that the rain of 7 and 8 July was not accompanied by high temperatures. The heavy rain and high temperature of 11 July was accompanied by the puzzlingly erratic behaviour of the glacier at 'West Camp' station, and by the second quicker horizontal movement of 'East Fan' station together with distinctly active forward movement of the fast moving 'East Camp' station. An even faster movement of this station on the evening of 10 July, however, coincides neither with rain nor high temperature. The high temperature, but without rain, of 16 July coincided with a slight but definite acceleration of both 'East Fan' and 'West Camp' stations, but with a retardation of 'East Camp' station.

The presence of deep, narrow gaps between the glacier and the side walls with the ice only in contact with protuberances of the rock walls, leads to the suggestion that erratic movements result from the melting of the ice which is pressing hard against these protuberances. The reality of these deep gaps was demonstrated by our accidentally dropping the container end of a 33 m tape down a gap that was almost invisible at the surface at 'West Camp' station. The tape rapidly unwound to its full length and then did not reach the bottom. It seems that the more uniform movement of the neighbouring ice farther from the side wall exerts a relatively uniform drag on the marginal ice, but that this latter ice jerks forward as it frees itself from the projections in the side wall, or from boulders jammed in the gap between the ice and the rock. Then after a large forward movement, a pause ensues until the drag is sufficient to overcome the side resistance once more. Some of the changes of velocity which relate neither to rain nor to temperature may be related to the varying resistance of the bulging side walls and to the encumbrance of boulders jammed between the ice and the solid rock.

GLACIAL EROSION

The forward surges, especially at 'West Camp' station, the jamming of the boulder at 'East End' station, the load of boulders always present in the gap between the glacier and the rock walls, and above all the grooved and striated appearance of the massive rock faces, help us to appreciate how glaciers may erode their beds. Immediately alongside 'East Fan' station, striations and grooves, some crossing each other, occur on a massive exposure of gneiss. Boulders and stones filled the gap between the rock mass and the ice just above the stake position. The grooves were remarkably fresh and unweathered, and the largest could almost hold the shaft of the ice axe. One ended abruptly at a steeply inclined quartz vein. The responsible stones and boulders seem to have ground almost away a bulge which ended abruptly on the downstream side at a quarried face. On the up-glacier side of this same rock mass, the surface was curved and generally smoother. This grooving and quarrying demonstrate the two chief ways in which glaciers remove massive rock with few joints and bedding planes. These effects have long been recognised but the cause has been little understood (Lewis, 1954).

McCall's (1960) approximations and calculations throw light on this problem. He followed others in discarding plucking as a mechanism because the yield stress in shear for ice is only 1 kg/cm², approximately 1/160 of that for granite. So in a tug-of-war the ice would break and not the granite. For grinding and grooving considerably greater forces need to be exerted on the rock walls. Consider an ideal case such as a cube of granite of 1 m side resting on the bed of a glacier. Compression on the up-glacier face can theoretically be 20,000 kg (i.e., 2 kg/cm², the yield stress for ice in compression, times 100 cm squared, the area of the surface). Shear along the top and side faces in line with the flow may theoretically equal 30,000 kg (i.e., 1 kg/cm² $\times$ 100 cm $\times$ 100 cm $\times$ 3). The ice may press up-glacier against the downstream face, it may not touch it at all, or it may exert a slight downstream drag, so we shall ignore this problematical and relatively small effect. So the maximum theoretical force may approximate to 50,000 kg, but even assuming it is only half this, it is still sufficient to shear off a projection with an initial surface of contact with the parent rock of 160 cm², an area greater than the area of your hand. Granite and other rocks are far stronger in compression than tension, yielding at about 1400 kg/cm². So if the cube pressed solely on one point it may be able to gouge an area of 18 cm². In order to appreciate the different possibilities for quarrying and for grooving, it is instructive to realise that this same block could have pulled away a piece of rock with a sound surface contact of about 180 cm². This may help to explain why glaciers can remove considerably more material from the downstream side of projections such as *roches moutonnées* than they can grind away from a bulging surface roughly parallel with the direction of flow.

There are usually lines or surfaces of weakness within rocks of which glacial quarrying will take full advantage. It is also necessary to remember that originally deep-seated igneous, and some other rocks may contain stresses within them which help the initial formation of such joint planes. So the effects of glacial quarrying on such rocks might be far greater than a simple calculation based on the relative strengths of ice and rock would suggest.

None of these processes of glacial erosion need be quick to accomplish prodigious erosion. If a glacier removed only 1 mm per year from its rock bed this would amount to 250 m in a quarter-million years, i.e. in about a quarter of the length of the Pleistocene Ice Age including both glacial and interglacial periods. 3 mm per year continuing for half the length of the Ice Age would accomplish 1500 m of erosion, probably sufficient to account for the quite exceptional over-deepening of the trough containing the Sogne Fjord. Such erosion requires a plentiful supply of boulders and stones at the surface of contact between ice and rock. Our experience on Austerdalsbreen and elsewhere suggests that there is no shortage of such grinding and prising tools, and they are constantly renewed by debris falling on to the glacier from the frost-shattered side walls far above. Perhaps the frequency of these tools along the sides of glaciers – witness the side moraines – may help glaciers to erode their side walls at a faster rate than their beds, a characteristic that is suggested by the rather wide U-shape of the typical glaciated valley. But the glaciers must slip actively along their side walls for this erosion to occur. Our present evidence suggests that this active side-slip occurs below ice falls where wide and deep rock basins are known to occur (E. de Martonne, 1911). However, the boulder that was jammed against the rock wall at 'East End' station and did not move could have accomplished no material erosion. This is because near the surface of a glacier, where hydrostatic pressure is low, the ice can flow round or over such a boulder, exerting little force upon it. The ice can only exert its full force if the boulder is covered by 60–70 m of ice.* The full force of the ice, limited only by its inherent strength, is then exerted on the boulder. At still greater depths the forces exerted by the moving ice may not be much greater owing to the limited strength of the ice. On the other hand the boulder may press very much harder against the bedrock especially if the surface of contact between rock and rock is free from ice or water. We need to know

* It seems likely that 60–70 ft was intended here by Lewis (see McCall, p. 254 of this volume, where a figure of 22 m, or about 70 ft, is given). C.E.

more about the mechanics of flow at these great depths to understand deep glacial erosion. One thing is fairly certain, that providing the glacier is slipping at all, velocity is a relatively minor factor, as the force exerted by the ice on a single boulder increases only as the quarter power of velocity (Glen, 1952). So velocity, which is so important for river erosion, may be less critical for glacial erosion. The main effect of greater velocity will be simply to increase the number of boulders passing over any given area of bedrock.

The availability of tools may also be a factor in bed as well as valley-side erosion by glaciers. They may never be as plentiful beneath a glacier as at the sides, but they may well be more plentiful in some areas than others. One of the few places where the bottom of a glacier has been properly studied was at the inner end of the lower tunnel dug through Vesl-Skautbreen (McCall, 1960). Angular blocks of rock occurred there, and the rock face showed signs both of grooving and quarrying. The bottom of the glacier was not, in fact, frozen hard on to the bedrock. The ice came away easily from the rock floor, perhaps because the heat from the Earth had melted the ice in intimate contact with the rock. Seeping meltwater would always tend to follow and enlarge such cracks. Such blocks of rock may also occur beneath Austerdalsbreen near the foot of the ice fall. Thinning and excessive crevassing in the swiftly moving ice falls may have allowed rocks tumbling from the cliffs on to the ice falls to penetrate deeply, and finally to sink slowly to the bottom because of their considerable mass and greater density than the surrounding, slowly yielding ice. This process would be accelerated by the fact that ice in the ice fall is already under considerable stress due to its rapid flow.

REFERENCES

BATTLE, W. R. B. (1952) 'Contributions to the glaciology of north-east Greenland, 1948–49, in Tyrolerdal and on Clavering Ø', *Meddelelser om Grønland*, CXXXVI 2, p. 28.

FORBES, J. D. (1842) 'Account of his recent observations on glaciers', *Edinburgh New Philosophical Journal*, XXXIII 338–52.

FORBES, J. D. (1843) *Travels through the Alps of Savoy* (Edinburgh: Simpkin) p. 141.

GLEN, J. W. (1952) 'Experiments on the deformation of ice', *Journal of Glaciology*, II 12, pp. 111–14.

GLEN, J. W. (1958) 'Measurement of the slip of a glacier past its side wall', *Journal of Glaciology*, III 23, pp. 188–93.

JACKSON, J. E., and THOMAS, E. (1960) 'Surveys and ice movements of Veslgjuvbreen', in W. V. Lewis (ed.), *Investigations on Norwegian Cirque Glaciers*

(London: Royal Geographical Society) chap. 6, pp. 63–7 (R.G.S. Research Series, no. 4).

LEWIS, W. V. (1954) 'Pressure release and glacial erosion', *Journal of Glaciology*, II 16, p. 422.

MCCALL, J. G. (1960) 'The flow characteristics of a cirque glacier and their effect on glacial structure and cirque formation', in W.V. Lewis (ed.), *Investigations on Norwegian Cirque Glaciers* (London: Royal Geographical Society) chap. 5, pp. 39–62 (R.G.S. Research Series, no. 4).

MARTONNE, E. DE (1911) 'L'érosion glaciaire et la formation des vallées alpines', *Annales de Geographie*, XIX 299.

MEIER, M. F. (1960) 'Mode of flow of Saskatchewan Glacier, Alberta, Canada', *U.S. Geological Survey, Professional Paper 351*, pp. 16–18.

NYE, J. F. (1952) 'The mechanics of glacier flow', *Journal of Glaciology*, II 12, pp. 82–93.

PILLEWIZER, W. (1950) 'Bewegungsstudien an Gletschern der Jostedalsbre in Südnorwegen', *Erdkunde*, IV, ht. 3/4, pp. 201–6.

WASHBURN, H. B., JR., and GOLDTHWAIT, R. P. (1937) 'Movement of South Crillon Glacier, Crillon Lake, Alaska', *Bulletin of the Geological Society of America*, XLVIII 11, pp. 1653–63.

WEERTMAN, J. (1957) 'On the sliding of glaciers', *Journal of Glaciology*, III 21, pp. 33–8.

11 The Flow Characteristics of a Cirque Glacier and their Effect on Cirque Formation

J. G. McCALL

GLACIATION is generally accepted as one of the primary factors in cirque formation. C. A. Cotton (1942) defines the processes of glaciation as erosion, transport and deposition of eroded debris. In order to understand better these processes a quantitative study was made of the flow characteristics and glacial structure of a small cirque glacier, Vesl-Skautbreen, Jotunheimen, Norway.

QUANTITATIVE OBSERVATIONS

The field-work was carried on in the summer and December of 1951 and in April and the summer of 1952. The aspects studied were: absolute glacier movements, relative or differential movements, tunnel closure, englacial temperatures, ice and firn densities and structures, and régime. These observations were concentrated for the most part along a vertical longitudinal section of the glacier (Fig. 11.1).

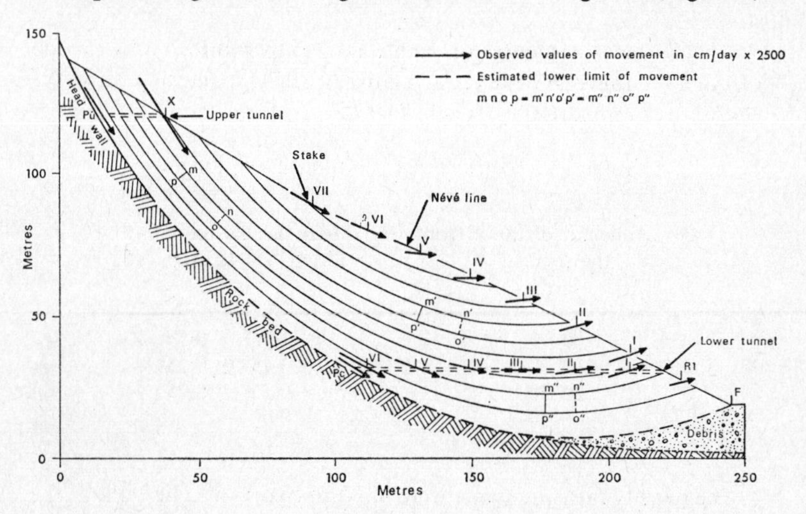

Fig. 11.1 Long section through Vesl-Skautbreen showing flow-lines and velocities

The absolute movements were obtained by a survey of surface stakes and tunnel pegs. In addition to the stakes above the tunnel, others were placed near the margins of the glacier (McCall, 1954, Fig. 1); all that need be said of them is that they indicated the usual marginal drag characteristic of all glaciers. A careful check was made to determine if any seasonal or short-term variations existed in the movement. All units of velocity are in cm/day (mean daily movement for the year August 1951 to August 1952). When expressed in vector form (Fig. 11.1), they are multiplied by a factor of 2500 in order to show on a diagram. In other words, all vectors represent approximately 7 years' movement relative to the scale of the diagram if conditions were to remain constant for that length of time. The results of all measurements can be summarised as follows: (*a*) all observed localities exhibited differential movement relative to a vertical line, (*b*) a higher point usually moved quicker than a lower point, (*c*) the differential movement was greatest in the basal zone of the glacier, and (*d*) at no place was there any evidence of differential movement along discrete planes as with thrust-faults, even along dirt and blue bands.

The preliminary results of tunnel closure have already been published (McCall, 1952). J. F. Nye (1953) has used them in his work on a flow law for ice. Briefly, they support the view that with ice the strain rate is an exponential (power of 4) function of the applied stress.

Rough measurements of englacial temperatures were made throughout the year; the data relative to the tongue region of the glacier are summarised in Table 11.1.

Table 11.1 Englacial temperatures in the tongue

Perpendicular distance from the surface in m	Mean annual temperature (Aug 1951 to Aug 1952) in °C
0–10	−1·5° ± 0·5°
10–20	−0·75° ± 0·25°
20–30	−0·25° ± 0·25°
30	0·0°

The density measurements were also approximate. The bulk of the glacier (all that below and down-glacier from an approximately horizontal line running from stake VI to the headwall (Fig. 11.1)) is ice of density 0·9. The firn of the glacier (above the horizontal line) had summer densities ranging from 0·6 at the surface to 0·8 at the depth of 18 m, as observed along the length of the upper tunnel. The relatively high surface value for the firn is attributed to wind-drifted snow, great amount of summer rainfall, and numerous avalanches from the headwall.

As regards régime, the glacier had a positive balance for the year August 1951 to August 1952; an increase of 4 m took place in the general elevation of the firn surface near the headwall, while the tongue thickened by an average amount of 0·3 m. Ablation during the summer of 1951 removed as much as 2 m of ice from the tongue compared with 1·3 m in 1952. The accumulation was of the 'wind-drift and avalanche' wedge-shaped type so characteristic of cirque glaciers (F. Matthes, 1900; G. R. Gibson and J. L. Dyson, 1939).

FLOW CHARACTERISTICS

In this paper, flow characteristics imply the speed and direction of ice movement within the glacier. Two techniques have been used in the past for illustrating these values: flow-line diagrams and velocity distribution ('velocity profile') diagrams.

Flow-lines. The first requisite for plotting the flow-lines was the shape of the lower limit (bounding line) of glacier movement. The fact that the glacier slides over its rock bed implies that the general shape of the bed must largely control this lower limit of movement. A reasonable estimation of the bed shape was obtained from the survey data (W. V. Lewis, 1949, p. 149; McCall, 1952, p. 122).

There is no doubt that the bedrock does not conform exactly to the regular curve (dashed line) shown in Fig. 11.1. However, the fact that ice does possess a degree of rigidity would lessen the possibility of its moving abruptly in and out of any small bed irregularity in a glacier of this thickness. The assumption is made, therefore, that the lower limit of movement approximates to the curve of the rock bed, except at the glacier terminus where the lower limit must rise from the bed so as to pass through stake F, whose movement was negligible. This rise towards F is attributed to the glacier over-riding its terminal moraine. It is discussed in the section on erosion.

The flow-lines were then plotted to comply with the shape of the lower limit of movement and with the observed velocities, expressed as vectors in Fig. 11.1. By definition (e.g. R. Koechlin, 1944, p. 131), the volume of material per unit width (assuming constant density) which flows between any two adjacent lines in a given time must be constant throughout. The flow-lines in Fig. 11.1 are based on this assumption. In the firn region the situation was complicated somewhat by the lesser densities, but the scale of the drawing is not large enough to make apparent any existing discrepancies. The implications of the spacing may be summarised briefly as follows: an increase in spacing or a divergence of flow-lines implies a decreasing velocity, and vice versa. It must be pointed out, however, that the slight convergence of flow-lines in the firn is the result of increasing density rather than increasing velocity.

The general flow-line pattern is in close agreement with that produced by H. F. Reid (1896). At the glacier surface, the direction of movement is down into the névé (accumulation), parallel at the névé-line, and out or up in the tongue (ablation). Reid (p. 917) explained the latter feature: 'In order that the general volume of the glacier should be preserved we must have below the névé-line, where there is melting, a component of the motion towards the surface, and this component is strongest where the melting is greatest. . . .' This was so at Vesl-Skautbreen, because the maximum component toward the surface occurred in the vicinity of the maximum ablation (near stakes I and II).

Reid (1896, p. 912) also pointed out, in his law of flow, that 'the greatest flow occurs through a section at the névé-line and diminishes as we go up or down the glacier from there; and, moreover, that the rate of diminution of flow becomes greater the farther we go from the névé-line'. In the law, the term 'flow' is used to denote quantity per unit of time or the velocity multiplied both by the sectional area and by the density. The maximum flow occurs through the section at the névé-line because the flow through any section above is lessened by that amount of accumulation between the higher section and the névé-line, and the flow through any section below is lessened by that amount of ablation between the lower section and the névé-line.

Velocity distribution. Whereas the flow-lines portray primarily the directions of movement, they give only an indirect impression of the velocity distribution. Lines of equal speed are therefore shown in Fig. 11.2 superimposed on the flow-line pattern. They were interpolated from the points of known velocity and represent lines on any one of which all points possess the same speed regardless of direction.

The absence of thrust-planes made possible the assumption that everywhere within the section there is a gradual transition between any two known velocities.

The plotting of the lines of equal speed is a somewhat conjectural procedure, so some qualification is necessary. Start with the points for which the mean daily movements for the year are known. These values range from 0·50–0·90 cm/day. An interval of 0·02 cm/day is used between the equal-speed lines, not to denote such great accuracy,

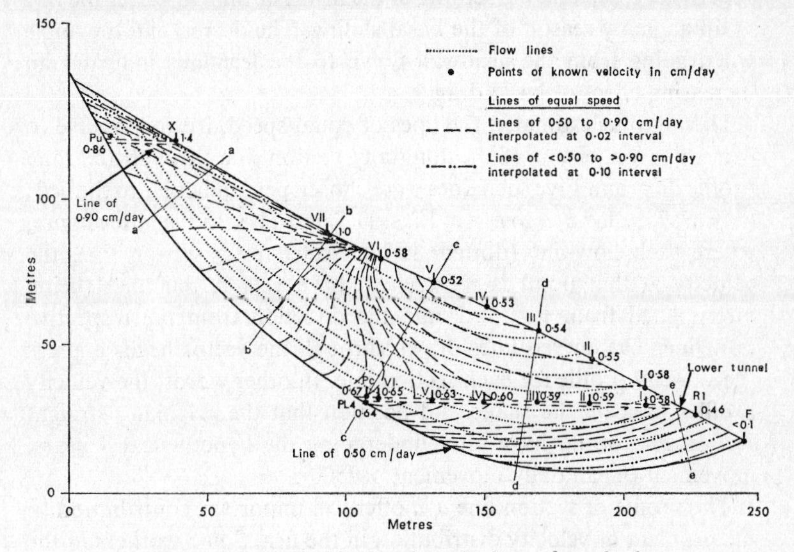

Fig. 11.2 Velocity distribution shown by lines of equal velocity

but to ensure sufficient lines being drawn to indicate the pattern of velocity distribution. For values greater than 0·90 and less than 0·50, the interval is increased to 0·1 because the speeds in those ranges are changing so rapidly that the smaller interval could not be shown on the diagram. For instance, there is a decrease from 0·46 at stake R_1 to <0·1 at stake F in a distance of only 20 m, and from 1·4 at stake X to 0·86 at peg Pu in a distance of only 25 m. The former decrease is the result of basal drag, while the latter is the result of the rapid compaction and down-glacier creep in the surface firn.

The general decrease in speed below the lower tunnel was based on the relative movement data which showed that in all tunnel alcoves a roof peg moved ahead of its floor peg (McCall, 1952, Fig. 3).

Beneath the rear of the tunnel the equal-speed lines are curved back sharply to signify localised differential movement which occurs there by reason of basal drag. On the other hand, the general decrease between the tunnel and the surface was based on the fact that the surface stakes have a lesser speed than the alcoves. The crowding of the lines between stakes VI and VII is caused by the rapid decrease in speed in going from the névé to the ice surface. The névé speed is greater because it is the sum of the general glacier movement and the superficial creep and the settling of the firn. The lines intersect the bed at an angle by reason of the basal sliding. The decrease in the angle when going from the headwall down to the terminus indicates an increasing effect of basal drag.

Having thus estimated the lines of equal speed, it was possible to determine velocity profiles along any section line through the long profile diagram. Five such lines were chosen perpendicular to the bed; they are labelled *a–a* to *e–e* in Figs. 11.2 and 11.3. In the latter figure, where each flow-line (dotted) is intersected by a section line, the velocity at that point is represented by a vector, the speed being interpolated from Fig. 11.2 and the direction from the respective flow-line. The curved lines drawn through the vector heads express the velocity profile for each section line. In other words, the velocity profiles indicate the shape and position that the originally straight section line would have after undergoing the hypothetical 7 years' movement (mean daily movement ×2500).

The profile of section line *a–a* offers an important contribution to the problem of velocity distribution in the firn. Some workers in the past (e.g. Matthes, 1949; R. Streiff-Becker, 1938–9) have assumed, after considering only surface movement, a profile of the nature shown by line *s–t* in Fig. 11.3. The discrepancy between it and that as estimated on Vesl-Skautbreen is obvious. The reason is that the surface zone velocities decrease rapidly in the first few metres of depth because of the relatively great rates of creep and compaction in the less dense superficial firn, as already pointed out. The effect would be greater still in winter when a snow cover exists and much care would have to be exercised when evaluating the overall movement of the firn from surface observations only.

In moving down the glacier to section line *b–b*, the profile has altered somewhat. Its new shape indicates that surface compaction has diminished and that basal drag has taken effect. The former is the result of a decreased thickness of firn, which in turn is the result of the wedge-shaped annual accumulation. The greater basal drag is due to increased depth and decreased bed slope.

The section line *c–c*, which is located below the névé-line and which therefore is free of firn effects, has two important features: (1) the surface velocity is slightly less than at depth, a maximum occurring at a distance of 5–10 m above the bed, and (2) the velocity vectors are not parallel, a maximum divergence of 23° existing between the surface and basal vectors. The features of line *c–c* are repeated in

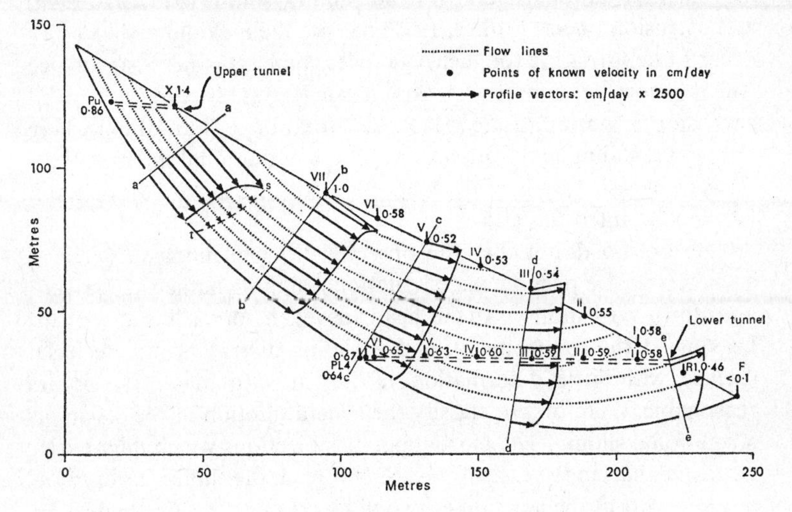

Fig. 11.3 Velocity distribution shown by velocity profiles

general by the next down-glacier section line *d–d*, but to a lesser extent. There the point of maximum velocity has moved up from the basal zone to a point approximately 25 m above the bed. The final profile *e–e* shows that the maximum velocity has returned to the surface and that a rapid decrease occurs systematically with depth. The inference is that the basal drag effect is distributed throughout the full depth of the relatively thin terminal ice.

A validity check on the velocity profiles was made by applying Reid's law of flow to the volumes represented by the area between each section line and its profile. In all cases, a reasonable agreement was obtained.

On the evidence of the velocity profiles, a locus of maximum velocity occurs at the surface in the névé, then drops abruptly into the glacier's interior at the névé-line, and remains at depth in the tongue only to return to the surface in the terminus. The position of the locus is in general agreement with that given by Matthes (1949), except that in the névé region where he placed his at depth, on Vesl-Skautbreen it was at the surface. The reasons for the velocity distribution in the névé have been discussed; it remains to explain the velocity distribution in the tongue.

Theoretical discussion. Both the current theories of cirque glacier movement, rotational sliding (Gibson and Dyson, 1939; Lewis, 1949) and extrusion flow (Matthes, 1949) require the maximum velocity to occur at depth in the tongue. A detailed analysis is necessary to see which theory best fits the observed conditions. In order to avoid confusion, it is appropriate to state that rotational sliding implies here only basal sliding and does not include the additional internal overthrusts suggested by Lewis (1949). Their absence at Vesl-Skautbreen has been pointed out above.

There is no doubt that the observed flow phenomena agree in general with the phenomena of rotational sliding, namely, a forward and downward motion at the head (névé), a forward and upward heaving at the toe (tongue), and a sliding along an arcuate basal surface with little deformation of the moving mass. It appears reasonable, therefore, to classify the general motion of the glacier as a rotational sliding. The most apparent objections which might arise would be that, in the case of Vesl-Skautbreen, the sliding is very slow and it occurs at the junction of two different materials (ice and rock), while a rotational landslide occurs suddenly along a definite surface of failure within a material.

An analysis of the observed-flow characteristics where extrusion flow is concerned must be based on the deformation or differential movement (creep) within the mass because the only tangible quality of the theory is that a glacier's interior is moving faster than its surface. An indirect approach has been made centring on the question: Are there any features of the deformation which cannot be accounted for by rotational sliding, but which can be explained by extrusion flow?

For the purpose of simplified dimensions, the long profile section of the glacier has been given a surface slope of 26° and an arcuate bed of a radius of 240 m. These dimensions were then adjusted in Fig. 11.4A so as to give a section which closely conforms to the observed section. If the mass in the figure is assumed to rotate as a rigid body, the flow-lines and velocity profiles would be as shown. These flow-lines are in general agreement with the observed lines (Fig. 11.1), except that they are everywhere parallel, while the observed lines are markedly divergent in the tongue region. In other words, since the flow-lines of rigid-body rotation are parallel, they do not account for the observed upward movements of stakes R to III. The glacier must,

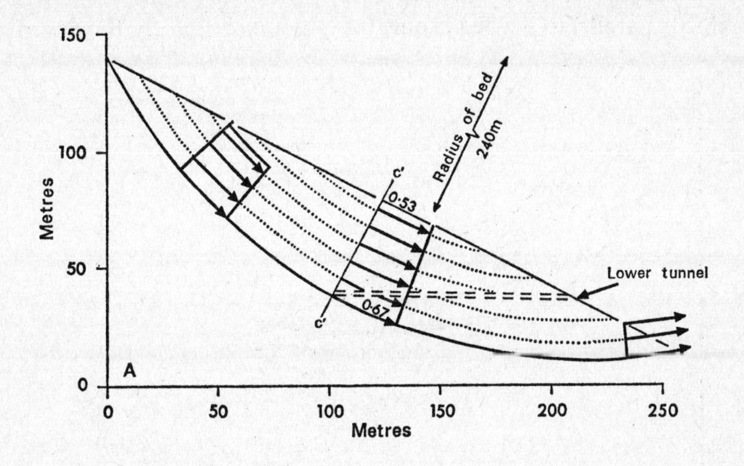

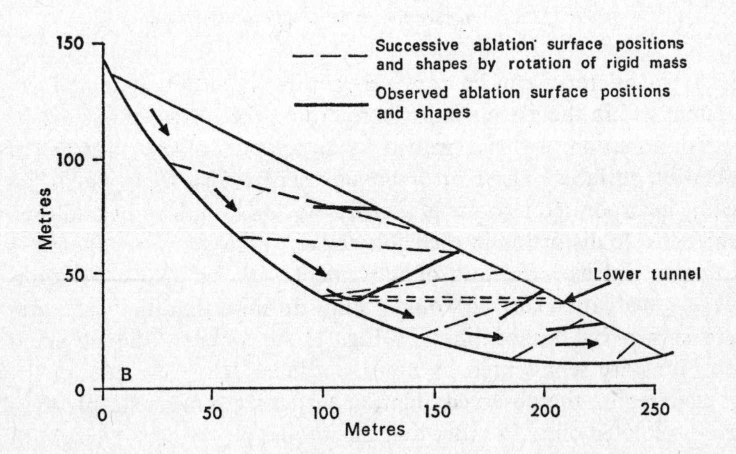

Fig. 11.4 A. *Velocity distribution for a rotating rigid mass;* B. *Ablation surfaces for a rotating rigid mass*

therefore, possess some inner stresses which cause the divergence and the upward movements.

In comparing velocity profiles, section *c′–c′* of the rigid body (Fig. 11.4A) corresponds to section *c–c* of the glacier (Fig. 11.3). If the basal vector in *c′–c′* be given the observed value of 0·67 cm/day, the surface vector must have a value of 0·53 by reason of its proportional radius of rotation. Yet in *c–c* the component of surface velocity parallel to the bed is only 0·48, or approximately 10 per cent

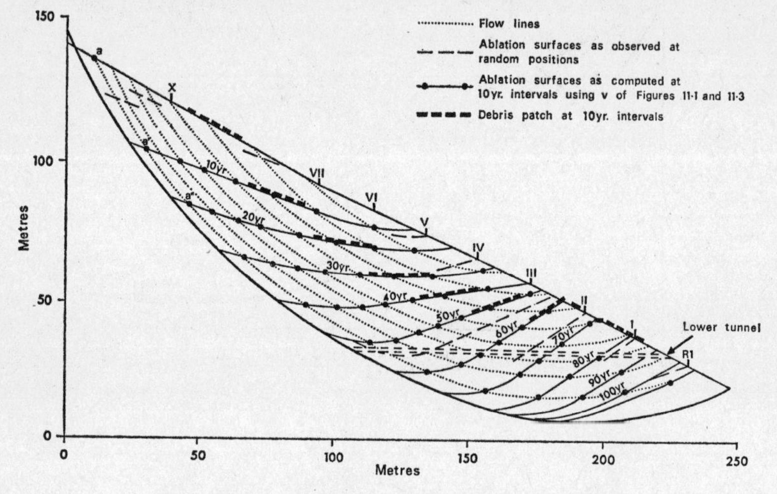

Fig. 11.5 Long section showing ablation surfaces

less. The inference can be drawn from this evidence that some force system within the glacier must be retarding the surface movement.

Confirmation of this is gained by an analysis of the positions of ablation surfaces. These are relic summer-névé surfaces which became incorporated into the glacier by the addition of snow and then subjected to distortion in their journey down-glacier. They therefore can be used as indicators of movement. The successive positions which an ablation surface would occupy during a rigid-body rotation are shown by dashed lines in Fig. 11.4B, whereas the observed positions are represented by unbroken lines. It is clear that in the tongue region the observed ablation surfaces dip more steeply than do the rotated ones. In order that this should be so, the movement of the outcropping ends of the observed layers must be retarded relative to the interior. It must be pointed out here that the flow-lines and the estimated velocity distribution of Figs. 11.1 and 11.2 give calculated layer positions which are in close agreement with the observed positions (Fig. 11.5).

The existence of surface retardation (R. Haefeli, 1940, p. 180) suggests that, in addition to rotational sliding, a form of 'extrusion flow' may be slightly active in the glacier movement. That is to say, the effect of the extrusion flow is superimposed upon the general effect of the rotational sliding because the surface zone is acting as a sloping dam, beneath and against which the interior ice is tending to flow. The critical qualities which the surface zone must possess in order to act as a dam for the interior ice are (1) that it should be less subject to creep (more rigid) than the interior ice, and (2) that its base should be braced firmly against some foundation. The fact that the winter and early summer temperatures in the surface zone are less than in the interior definitely favours a more rigid condition (J. W. Glen, 1953) near the surface than at depth, and an ideal foundation is offered by the terminal moraine which in itself forms the lower portion of the dam. The dam presents only a slight obstruction to the glacier movement and is undergoing a continuous outward deformation toward a bulged shape, a condition of the tongue surface which was very obvious to the eye at the site. It is because of this form of 'dam failure' that this rather specialised form of 'extrusion flow' phenomenon may occur. The glacier's interior must supply a greater amount of ice to the point of maximum bulging (in the vicinity of stake I) and, in doing so, the flow at depth beneath the upper end of the dam (in the vicinity of stakes V and VI) is increased, relatively speaking.

ICE–ROCK CONTACT

The ice–rock contact at the junction of the glacier with its bed differed greatly at the ends of the two tunnels. Under the névé, at the end of the upper tunnel, a large headwall gap (Lewis, 1953) was encountered between the firn and the headwall, whereas under the tongue, at the end of the lower tunnel, the ice was in closer contact with its rock bed.

The general position of the headwall gap can be seen in Fig. 11.1, and a sketch section of it is given in Fig. 11.6. Its dimensions cannot be given exactly for it was most irregular and not fully explored. Roughly speaking, it had a depth of 50 m, a width of 1–2 m, and it gave the impression of extending laterally over much of the breadth of the headwall. No light entered the gap, but openings to the small randkluft above must have existed for the tunnel caused a strong circulation of air. This circulation was always down and out through the tunnel, where its strength was sufficient to blow out a candle flame.

Throughout August, both in 1951 and 1952, the headwall side of the gap was covered in most places by a coating of clear ice which was caused by the freezing of melt- and rainwater coming in from above. The firn side (roof) was supported in different places by irregular projections of the headwall. As the roof slid over these projections, grooves up to 5 cm deep were made in it, thus giving it a corrugated (mullion) appearance. Some of the corrugations ran free from the upper limit of the gap to the lower limit, a distance of roughly 50 m. Since the roof's movement is of the order of 3 m/yr, a point on such a corrugation must have remained free of the headwall for an interval of at least 15 years. Thus the gap is not an ephemeral feature but a fairly permanent one.

The mechanics of the gap's formation are undoubtedly related to those of bergschrunds and randklufts. Measurements of the roof's movement showed that its direction coincides with the roof slope, approximately 57°, and the gap exists because the headwall has a greater slope than the direction of movement. The gap is eventually closed at depth when the weight of the overlying firn becomes sufficient to force the movement down against the headwall. The gap's upper limit, between itself and the randkluft, is closed because of the 'cantilevered' nature of the firn above requiring support from the headwall. If the glacier's forward movement was more pronounced, the upper end would pull away from the headwall and the gap would become part of the randkluft, or even develop into a bergschrund. W. D. Johnson's (1904) famous bergschrund may well have been a gap which was open at its upper limit, for he describes a great depth (100 ft) of exposed headwall. Normally only at the bottom of a bergschrund is the headwall or rock bed exposed. On the other hand, the 'schrundless' cirque glaciers found by I. Bowman (1920, p. 295) in the Andes may also possess headwall gaps which, being closed at their upper limits, are not visible.

In contrast with the névé conditions, the ice of the tongue, at the end of the lower tunnel, was generally in close contact with its bed. The few small gaps which did exist were on the lee side of convex irregularities in the bed and were only 1 cm or less in thickness. This indicated that they are also the result of glacier movement; that is, the ice after passing over a bed irregularity is not forced immediately back to the bed because of insufficient pressure. T. G. Bonney (1876) observed such small gaps in a natural grotto in an alpine glacier, while larger ones have been reported by H. Carol (1947) on the lee side of roches moutonnées and by Haefeli (1951) beneath an icefall (see also Lewis, 1954).

The most outstanding feature of this contact was the sole, a debris-filled layer between the glacier ice and the bed (Fig. 11.6, this paper, and McCall, 1952, Fig. 5). Macroscopically, the ice matrix of the sole was distinctly different from the glacier ice above; it was transparent, bubble-free, layer-free and full of rock debris ranging in size from fines to boulders. Because of these qualities, such ice must have formed directly from the freezing of water and not from compaction of firn, as does the glacier ice. The sole had a depth of roughly 30 cm and probably originated as follows. Some of the water running down the headwall finds its way to the bottom of the headwall gap where it freezes. This freezing incorporates the fragments of rock debris lodged there (Fig. 11.6) and the whole of the frozen mass is then pulled along in a continuous manner by the glacier movement, thus forming the sole. Other fragments may be incorporated into the sole by corrasion of the bed (see next section).

What becomes of the sole at the terminus is not definitely known, for that vicinity was covered by snow; a pit dug there revealed no outcrop, but it may not have been deep enough. However, the velocity distribution (Fig. 11.2) suggests that the movement of the sole gradually dies out as the terminus is approached, which implies that the debris of the sole is being deposited under the terminus.

CIRQUE FORMATION

The impressions gained from Vesl-Skautbreen concerning cirque formation agree with many of the generally accepted views (e.g. Cotton, 1942, pp. 169–88); specifically, the role of glaciation is thought to be very important. The ensuing discussion of the glaciation is grouped under its three processes: sapping, corrasion and transport.

The term sapping, as used here, implies frost-riving on the rock slopes under the margins of a glacier. It is produced by the freezing of any water which flows in under the 'cold' glacier and, in the case of cirques, it results in a horizontal retreat of the headwall. Corrasion includes abrasion and plucking of a glacier's bed and is produced by the glacier's movement. Its action is primarily in a vertical sense; that is, it tends mainly to deepen a cirque. Transport, the third process of glaciation, involves the removal of eroded debris. It is important in

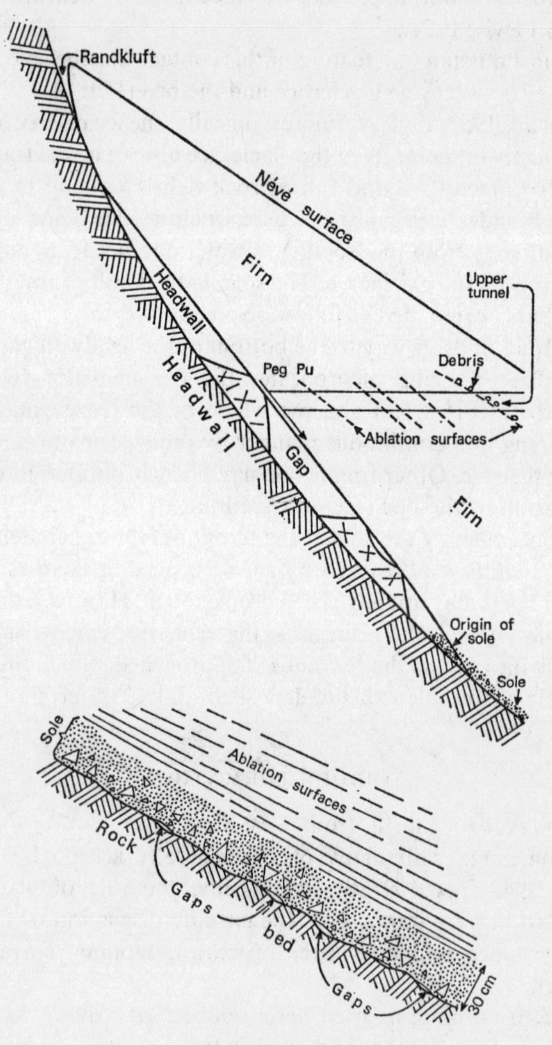

Fig. 11.6 Ice–rock contact under the névé (upper tunnel) and under the glacier (lower tunnel)

that its clearing away of debris offers continuously fresh working surfaces to the processes of sapping and corrasion.

Sapping. Ideal conditions for sapping were offered by the headwall gap (Lewis, 1953). The resemblance between this gap and W. D. Johnson's bergschrund was pointed out in the preceding section. In fact, the resemblance is so pronounced that a portion of his (Johnson, 1904, p. 574) description fits the conditions in the gap, word for word: 'The glacier side of the crevasse presented the more clearly defined wall. The rock face, though hard and undecayed, was much riven, its fracture planes outlining sharply angular masses in all stages of displacement and dislodgement. Several blocks were tipped forward and rested against the opposite wall of ice; others, quite removed across the gap, were incorporated in the glacier mass at its base. Icicles of great size, and stalagmitic masses were abundant; the fallen blocks in the large part were ice-sheeted; and open seams in the cliff held films of this clear ice.'

He went on to suggest that a freeze–thaw action was favoured by a diurnal fluctuation in air temperature (in the schrund) above and below the freezing-point. The later investigations of Battle (Battle and Lewis, 1951) showed that such a fluctuation was not present in other bergschrunds where the temperatures remained for the most part below freezing. These conditions do not exclude sapping, for, as Lewis (1938, 1940) and L. H. McCabe (1939) pointed out, a freeze–thaw action can be promoted by meltwater which descends into a schrund and which, on coming to rest, freezes because of the sub-zero temperatures.

At Vesl-Skautbreen there was an abundant supply of water into the headwall gap. In summer practically all the rain which fell on the headwall ran down into the gap and, during severe rainfalls, it was observed at various places in the gap that the incoming water appeared to enter through small tunnels. When rain was not falling, a steady supply of water was furnished by the melting of the large cornice and snow patches on the top of the headwall. Some of this 'summer' water collected in pools which were observed in the bottom of the gap in August 1951 and 1952. Such a condition may be what led T. C. Chamberlin and R. T. Chamberlin (1911, p. 212) to write, 'the exit of . . . internal drainage takes place more largely at or near the foot of the cirque wall than at a higher level (and thus promotes sapping at the foot)'.

The supply of water was not limited entirely to the summer months because even in December and April rain fell in the cirque. Also, when the air temperature was only slightly below freezing, as was often the case, any snow which fell on the steep, bare headwall melted immediately and its meltwater ran down into the gap.

The measurements made of the air temperature in the gap in August 1951 showed that it was −0·6°C a few hours after the completion of the upper tunnel. All later measurements (both in August 1951 and August 1952) showed the temperature to be 0°C. This was undoubtedly due in part to the unnatural air circulation caused by the tunnel, as pointed out above. No measurements were made in December and April because deep snow prevented entry into the tunnel, but it seems reasonable to assume that the temperatures in the gap at those times were of the same general order as the exterior air temperatures, whose daily means were below freezing. The cold air could enter the gap since both the glacier's movement and the incoming water would tend to keep its upper limit open in places, as was the case in summer. Thus in winter even the large pools of 'summer' water should freeze. This could cause a more severe sapping at the bottom of the gap and may help to explain the steepness of the headwall. Also it might be responsible for the marked breaks-of-slope or 'schrundlines' at the foot of some cirque headwalls. These have been described by G. K. Gilbert (1904) in his work in the Sierra Mountains of California, and by Lewis (1938, 1940) in his studies on glacier-free cirques in Great Britain.

The sapping at Vesl-Skautbreen appeared to be greatly facilitated by the well-developed joints in the rock of the headwall. This was shown by the general lack of any further frost action on the 'sapped' blocks of rock which formed the moraine and screes in the area. That is, after the 'joint-free' blocks were removed from their 'in situ' positions, further frost action on them was practically non-existent.

The slope of the headwall (approximately 60°) appeared to be related to the joints. Its face was made up of huge slabs of rock which often overhung on their lower sides – a condition which suggests that the sapping tended at times to undercut the headwall. If this is so, an overall retreat of the headwall can be imagined. D. W. Johnson (1941, p. 261) wrote: 'If sapping at the bergschrund zone can explain retreat of the base of the headwall, retreat of the entire headwall, however high, is automatically accounted for.'

Additional evidence for sapping was shown by the heavy concentration of debris which was being moved in the sole of the glacier

(McCall, 1952, Fig. 5). The bulk of this debris is believed to be derived from the 'sapped' fragments in the headwall gap, and if the concentration is typical of the whole of the glacier's bottom, an appreciable amount of sapping occurred during at least the past few decades. This continuous removal and transport of debris by the glacier provides a fresh surface for the sapping to work on. Cotton (1942, p. 176) emphasised the importance of this when he wrote: 'Working alone in the open air such a process (sapping) would soon slow down to zero pace as a protective layer of talus material buried the bedrock surface; but in association with a névé, which has the flowing motion of glacier ice, and which must in some way drag out and carry away frost-riven blocks of rock, continuity of the process of sapping by frost-riving can be understood.'

Corrasion. Of the two processes of abrasion and plucking (quarrying) which make up corrasion, abrasion appears to be the more active at the present time at the end of the lower tunnel through Vesl-Skautbreen. The evidence is twofold: the presence of rock flour on the surface of the rock bed at the rear of the lower tunnel, and the agreement between the directions of the parallel striations on the bed and the direction of glacier movement in the vicinity. In all, approximately 6 m^2 of the solid bed was exposed to view and its surface was mostly smoothed, but not highly polished, and apparently free of joints. There were some angularities in the surface, which suggest block removal. This could be interpreted as evidence of plucking.

It is extremely difficult, however, to imagine how the ice could possibly pluck anything more from this surface. In order that plucking should take place, the initial requirement would be that there should be a very strong adherence between the ice and the bed; E. Ljunger (1930) and O. D. von Engeln (1937) in their considerations of plucking have assumed that this is so. However, when the ice was being cleared from the bed of Vesl-Skautbreen, no such evidence was found; the ice came cleanly and easily away from the bed.

Even if adherence occurred, the relatively small yield stress in shear (1 kg/cm^2, see below) for ice compared with that of a rock such as gabbro ($\sim$160 kg/cm^2, see below) would prevent the ice from plucking anything but previously loosened fragments. Thus, as suggested by M. Demorest (1939), the existence of loosened or highly jointed material is essential for effective plucking.

Some writers consider sapping to be in the category of plucking, but even so the blocks are loosened in the first place by frost-riving and not by the 'pluck' of the glacier. There was no evidence of frost action on the observed portion of the bed; its surface remained damp and unfrozen throughout the year.

An essential condition for abrasion is the presence of rock debris (cutting tools) in the glacier's sole; the fact that clear ice cannot abrade, but rather is abraded, was emphasised by I. C. Russell (1895). This was shown at Vesl-Skautbreen by the grooves in the ice roof of the headwall gap and on the underside of the sole. The abrasion of the rock bed takes place because a constant supply of cutting tools is furnished to the sole by the sapping of the headwall (Fig. 11.6).

The mechanical action of abrasion can be of two types: (1) a planing or shearing-off of any small protruding irregularities of the bed, and (2) a gouging or crushing-out of striations. It can therefore be reasoned that, for the former, the critical forces on a fragmental cutting tool are those which act parallel to the bed and tend to move the fragment along the bed. For the latter the critical forces are those in a vertical direction which cause the fragment to bear on the bed. These forces will depend upon the strength of the ice and will be independent of the rate of flow. Investigations by M. Waeber (1942–3), R. Haefeli and P. Kasser (1951) and J. F. Nye (1952) all suggest that the yield stress in compression for ice at 0°C is of the order of 2 kg/cm^2. The calculations by Nye (1952) for the basal shear stress of various glaciers and by the author for Vesl-Skautbreen suggest the value of 1 kg/cm^2 for the approximate yield stress in shear. If the assumption be made that these stresses are constant throughout a glacier regardless of depth, the above values will govern the forces which a glacier can exert on a fragment.

It is obvious that there will then be a limit to the abrasive effectiveness of a fragment and that the limit will depend directly upon the effective area of the fragment. The larger the fragment is, the larger the limit. Once this limit is reached, no further increase in the force can be developed on the fragment even though the glacier may become larger or deeper.

In order to get some idea of the force which can be developed, a block with the idealised dimension of $1 \times 1 \times 1$ m will be considered. If this block is resting flat on a horizontal bed with one of its faces normal to the direction of glacier flow, the theoretical maximum horizontal force, parallel to the bed, exerted by the glacier against it would be:

Compression on the up-glacier vertical face normal to flow,

$$(2\ \mathrm{kg/cm^2})(100\ \mathrm{cm})(100\ \mathrm{cm}) = 20{,}000\ \mathrm{kg}$$

Shear along the top and side faces in line with flow,

$$(1\ \mathrm{kg/cm^2})(100\ \mathrm{cm})(100\ \mathrm{cm})(3) = 30{,}000\ \mathrm{kg}$$

In the calculation the assumptions are made that (1) the ice adheres firmly to the top and sides, and (2) a gap exists between the ice and the down-glacier vertical face. Since these assumptions are not likely to occur, the chances of the 50,000-kg force ever developing fully seem slight. However, even if the force can be reduced by 50 per cent, the remaining 25,000 kg is a force large enough to crush or to plane off any small irregularities on either the bed or the block.

For example, if the rock be given a compressive strength of 1400 kg/cm² and a shear strength of 160 kg/cm², the bearing surface between the block and the obstructing irregularity would have to be greater than

$$\frac{25{,}000\ \mathrm{kg}}{1400\ \mathrm{kg/cm^2}} \quad \text{or} \quad 18\ \mathrm{cm^2}$$

to prevent crushing of either the block or the irregularity. Similarly, the area of a horizontal cross-section through the irregularity would have to be greater than

$$\frac{25{,}000\ \mathrm{kg}}{160\ \mathrm{kg/cm^2}} \quad \text{or} \quad 160\ \mathrm{cm^2}$$

to prevent a shearing-off of the irregularity.

In the case of gouging-out striations, the theoretical maximum vertical force which could be developed between the block and the bed would be:

Weight of the block, assuming a density of 3,

$$(3)(1000\ \mathrm{kg}) = 3000\ \mathrm{kg}$$

Compression on the top face,

$$(2\ \mathrm{kg/cm^2})(100\ \mathrm{cm})(100\ \mathrm{cm}) = 20{,}000\ \mathrm{kg}$$

for a total of 23,000 kg.

The shearing forces along the vertical faces would not be effective since the ice and the block are resting on the bed. Gouging or crushing could then occur if the bearing area between the block and the bed is less than

$$\frac{23{,}000\ \text{kg}}{1400\ \text{kg/cm}^2} \quad \text{or} \quad 16\ \text{cm}^2$$

The depth of ice required to give the vertical load of 2 kg/cm^2 is 22 m. Any depth greater than this would cause the ice to flow around and under the block and the bearing of the block on the bed would not be increased. Thus, the amount of abrasion is directly proportional to the effective area of a fragment and is independent of the glacier depth at depths greater than 22 m. In other words, given the same bed conditions and ice temperatures and the same type of cutting tools, an extremely deep glacier would not cause any more abrasion than would a thin glacier whose depth was greater than 22 m.

These analyses are complicated by the possibility that a block might roll over the obstructing irregularity. An estimate of the shear stress developed on the faces of such a 'rolling' block would be that the force required to revolve the $1 \times 1 \times 1$ m block through the ice is of the order of the theoretical maximum horizontal force. Thus, even though the block rolled over, an appreciable force would still be acting against the obstruction.

In the event of the obstruction being large enough to prevent the block from rolling over and strong enough to resist its push, the ice would flow around the block. Other fragments would then be brought to bear against it. Eventually, this accumulation of debris might become large enough to provide the ice with a sufficient surface to act on, and the obstruction would be destroyed; or, as the debris accumulates, it might expand its 'anchorage to the bed' to other nearby obstructions and the ice would be forced to flow over the entire mass of debris. I. C. Russell (1895) has suggested that this is how some drumlins originate.

The situation at the terminus of Vesl-Skautbreen is somewhat similar. Because the sole was continuously bringing debris to the terminus and because there was no clear evidence that this debris was being discharged from under the glacier, the fragments must be slowly accumulating under the terminus. The glacier moves up and over these fragments (Fig. 11.1) because their motion is arrested against the large terminal moraine. This overriding of terminal moraines has been discussed by Gibson and Dyson (1939), in particular, and by many other writers. Because of overriding, there can be little or no corrasion of the bed under the terminus of Vesl-Skautbreen. This condition, together with the active corrasion up-glacier from the terminus, could very well lead to the formation of the basins so often found in cirques; and since the corrasion is the result of the movement of a glacier, it seems reasonable to assume that the shape of the basin will tend to conform to the shape of the flow-lines. Lewis (1949, p. 156) has described the phenomenon: 'When a section of a glacier is moving in a rotational slip the bedrock beneath must be moulded into a basin fitted to this rotational movement.' In a later paper, he (Clark and Lewis, 1951, p. 552) went on to say: 'We . . . think that the tendency to rotate is inherent in certain glaciers and that this may, under favourable circumstances, lead to the scouring-out of basins in the rock floor beneath. True, once such a basin is formed, the tendency for rotation to occur is much increased.'

The impression gained at Vesl-Skautbreen, however, is that this corrasive action is far less effective at present in the overall enlargement of the cirque than the sapping of the headwall. If this situation is typical of past conditions, the cirque has been enlarged primarily in a horizontal or headward manner by sapping. This idea has been advocated for other cirques by J. W. Evans (1913), G. Taylor (1914, 1922) and Lewis (1938, 1940). It has been emphasised recently by M. Boyé (1952, p. 20), who wrote: 'It [snowpatch or glacier] only supports a periglacial micro-climate favourable to frost action, which seems to be . . . the main morphogenic factor.'

The frost action is greatly facilitated by the geological structure in the Jotunheim area. The importance of the joints in the rock of the headwall has been pointed out; their importance to the general area was also evident. Many protruding outcrops of peridotite occur throughout the area and in most cases they show a much more advanced state of weathering than do the surrounding fragments of gneiss. These outcrops stand out above the gneiss presumably because the peridotites, possessing few or no joints in the unweathered state, were more resistant to direct glacial erosion. In some places the outcrops extend across the path of the former glacier movement and have caused basins to form on their up-glacier sides. This was especially evident in Glitterholet and suggests that the geological structure as well as glacier flow characteristics can contribute to the formation of basins.

TRANSPORT

The importance of debris transport in glacial erosion has been emphasised by many writers. In brief, this process clears away the debris and thereby prevents it from forming scree-slopes which would protect the bounding walls of a glacier from further frost action.

At Vesl-Skautbreen the transported debris can be divided into three groups:

(1) *The basal fragments* (*in the sole*). These are derived primarily from the headwall gap (by sapping) and secondarily from the bed (by corrasion) and from the portion of the headwall above the glacier from which some frost-riven fragments fall into the randkluft. All this basal debris is eventually deposited under the terminus.

(2) *The englacial fragments*. These are derived wholly from the portion of the headwall above the glacier. They fall on to the névé and become incorporated into the glacier (Fig. 11.5). When they eventually emerge at the surface of the tongue they slide or roll down to the terminus.

(3) *The supraglacial fragments*. These also are derived from the portion of the headwall above the glacier. They fall on to the névé and glissade or roll down the névé and tongue surfaces to the terminus.

Since all three groups contribute to the terminal moraine, the large size of this in comparison to the small glacier can be understood – the moraine is composed of debris derived from both glacial and subaerial erosion. The impression was gained that the latter is now the chief contributor, and because of this an appreciable portion of the moraine is in reality an accumulation of talus or pro-talus.

CONCLUSION

In conclusion, the more important inferences concerning the activities of the cirque glacier Vesl-Skautbreen are recapitulated below:

(1) The general glacier movement is analogous to that of a rotational landslide.

(2) A slight form of extrusion flow can be interpreted as being superimposed on the general rotational movement.

(3) The glacier bands mark the outcrop of accumulation layers and ablation surfaces which originated in the névé; the shape of the layers subsequent to their deposition is the direct result of the flow characteristics.

(4) Rather than a well-developed randkluft or bergschrund, the glacier possessed a deep headwall gap which was not visible from the surface. Such a gap may be a feature common to many 'schrundless' and other cirque glaciers.

(5) The debris on the tongue surface results from rockfalls down the headwall. There were no indications that any surface debris had been brought up from the bed.

(6) The cirque is being enlarged horizontally by the sapping of the headwall. The sapping is helped by (*a*) the headwall gap, (*b*) the freezing conditions maintained therein by the presence of the glacier, and (*c*) the jointed structure of the rock.

(7) Corrasion of the bed, at the present time, results primarily from abrasion; but it appears to be a much less effective erosive agent than the sapping.

(8) Calculations involving the strengths of rock and ice give a satisfactory explanation of the mechanics of abrasion.

(9) The rotational movement of the glacier favours the formation of a rock basin.

(10) The large terminal moraine is composed of a glacial moraine capped by a relatively great amount of talus.

REFERENCES

BATTLE, W. R. B., and LEWIS, W. V. (1951) 'Temperature observations in bergschrunds', *J. Geol.*, LIX 537–45.

BONNEY, T. G. (1876) 'Some notes on glaciers', *Geol. Mag.*, III 197–9.

BOWMAN, I. (1920) *The Andes of Southern Peru* (London: Constable).

BOYÉ, M. (1952) 'Névés et érosion glaciaires', *Rev. Géomorph. dyn.*, I 20–36.

CAROL, H. (1947) 'Formation of roches moutonnées', *J. Glaciol.*, I 57–9.

CHAMBERLIN, T. C., and CHAMBERLIN, R. T. (1911) 'Certain phases of glacial erosion', *J. Geol.*, XIX 193–216.

CLARK, J. M., and LEWIS, W. V. (1951) 'Rotational movement in cirque and valley glaciers', *J. Geol.*, LIX 546–66.

COTTON, C. A. (1942) *Climatic Accidents* (Christchurch, N.Z.: Whitcombe & Tombs).

DEMOREST, M. (1939) 'Glacial movement and erosion', *Am. J. Sci.*, CCXXXVII 594–605.

EVANS, J. W. (1913) 'The wearing down of rocks', *Proc. Geol. Ass., Lond.*, XXIV 241–300.
GIBSON, G. R., and DYSON, J. L. (1939) 'Grinnell Glacier', *Bull. geol. Soc. Am.*, L 681–96.
GILBERT, G. K. (1904) 'Crest lines in the High Sierras', *J. Geol.*, XII 579–88.
GLEN, J. W. (1953) 'Rate of flow of polycrystalline ice', *Nature*, CLXXII 721.
HAEFELI, R. (1940) 'Zur Mechanik aussergewöhnlicher Gletscherschwankungen', *Schweiz. Bauztg.*, CXV 179–84.
—— (1951) 'Some observations on glacier flow', *J. Glaciol.*, I 496–8.
—— and KASSER, P. (1951) 'Geschwindigkeitverhältnisse und Verformungen in einem Eisstollen', *Gen. Ass. int. Un. Géod., Brussels*, I 222–36.
JOHNSON, D. W. (1941) 'The function of meltwater in cirque formation', *J. Geomorph.*, IV 253–62.
JOHNSON, W. D. (1904) 'The profile of maturity in alpine glacial erosion', *J. Geol.*, XII 569–78.
KOECHLIN, R. (1944) *Les glaciers et leur mécanisme* (Lausanne: Rouge).
LEWIS, W. V. (1938) 'A meltwater hypothesis of cirque formation', *Geol. Mag.*, LXXV 249–65.
—— (1940) 'The function of meltwater in cirque formation', *Geogr. Rev.*, XXX 64–83.
—— (1949) 'Glacial movement by rotational slipping', in *Glaciers and Climate*, *Geogr. Ann., Stockh.*, XXXI 146–58.
—— (1953) 'Tunnel through a glacier', *The Times Sci. Rev.*, IX 10–13.
—— (1954) 'Pressure release and glacial erosion', *J. Glaciol.*, II 417–22.
LJUNGER, E. (1930) 'Spaltentektonik und Morphologie der schwedischen Skagerrack-Küste', *Bull. geol. Instn. Univ. Uppsala*, XXI 1–478.
MCCABE, L. H. (1939) 'Nivation and corrie erosion in W. Spitsbergen', *Geogr. J.*, XCIV 447–65.
MCCALL, J. G. (1952) 'The internal structure of a cirque glacier', *J. Glaciol.*, II 122–30.
—— (1954) 'Glacial tunnelling and related observations', *Polar Rec.*, VII 120–36.
MATTHES, F. E. (1900) 'Glacial sculpture of the Bighorn Mts., Wyoming', *U.S. geol. Surv. Ann. Rept.*, XXI 167–90.
—— (1949) 'Glaciers', in O. E. Meinzer (ed.), *Physics of the Earth*, vol. IX: *Hydrology*, pp. 149–220 (New York: McGraw-Hill).
NYE, J. F. (1952) 'The mechanics of glacier flow', *J. Glaciol.*, II 82–93.
—— (1953) 'The flow law of ice', *Proc. R. Soc. Lond.*, ser. A, CCXIX 477–89.
REID, H. F. (1896) 'The mechanics of glaciers', *J. Geol.*, IV 912–28.
RUSSELL, I. C. (1895) 'Influence of debris on the flow of glaciers', *J. Geol.*, III 823–32.
STREIFF-BECKER, R. (1938–9) 'Zur Dynamik des Firneises', *Z. Gletscherk., Berl.*, XXVI 1–21.
TAYLOR, G. (1914), 'Physiography and glacial geology of east Antarctic', *Geogr. J.*, XLIV 365–82, 452–67, 553–71.
—— (1922) 'The physiography of the McMurdo Sound and Granite Harbour region', in *British Antarctic Expedition 1910–13* (London: Harrison).
VON ENGELN, O. D. (1937) 'Rock sculpture by glaciers', *Geogr. Rev.*, XXVII 478–82.
WAEBER, M. (1942–3) 'Observations faites au glacier de Tré-la-tête', *Rev. Géogr. alp.*, XXX/XXXI 319–43.

12 Direct Observation of the Mechanism of Glacier Sliding over Bedrock

BARCLAY KAMB and
E. LaCHAPELLE

It has been demonstrated by direct observation in tunnels and by measurements of ice deformation in boreholes that an important mechanism in the flow of temperate glaciers is sliding of the ice over the bedrock at its base. Basal sliding or slip (R. P. Sharp, 1954, p. 826) normally accounts for the order of 50 per cent of the total surface velocity of typical valley glaciers in their thicker parts, the remainder being due to internal deformation within the ice (J. A. F. Gerrard, M. F. Perutz and A. Roch, 1952; J. G. McCall, 1952; W. H. Mathews, 1959; R. L. Shreve, 1961; J. C. Savage and W. S. B. Paterson, 1963).

Observations of internal deformation are by now relatively numerous, and their broad features can be rather well accounted for quantitatively or semi-quantitatively, in terms of experimentally or theoretically derived concepts. Up to now, however, few direct observations of the basal-slip process in operation have been reported. Marginal slip has been measured on several glaciers (Sharp, 1954, p. 826; J. W. Glen, 1961), but the conditions of surface observation are such that it is difficult to draw conclusions as to the mechanism appropriate to basal slip at depth. Observations of basal slip at depths of 20–50 m in ice tunnels and in deep marginal crevasses have been reported by H. Carol (1947), R. Haefeli (1951) and McCall (1952); in none of these cases, however, was the lowermost ice investigated closely enough to elucidate the process and rate-controlling mechanism of basal slip, attention being instead focused on other questions, such as erosion of the bed.

In a theoretical treatment of the basal-slip mechanism, given by J. Weertman (1957, 1962), the phenomenon is attributed very plausibly to the combined operation of two processes: (1) regelation-slip, involving melting of the basal ice at points of increased pressure and refreezing at points of decreased pressure; (2) plastic flow, involving deformation of the ice due to stress concentrations. The regelation-slip process provides the necessary physical decoupling of the ice from its bed; the associated plastic flow is in a sense secondary, in that it simply allows the slipping ice mass to accommodate more readily to the larger obstacles in its path. It is known from tunnel observations in Greenland (personal communication from H. Bader) that, when the temperature at the bed is below freezing, regelation-slip cannot occur.

Regelation-slip is a process which intimately combines mass transport and heat transport at the glacier sole. It involves: (1) heat transport from points of local freezing to points of local melting; (2) mass transport of liquid water, in a thin basal layer, from points of melting to points of refreezing; (3) bulk transport of the main ice mass above, resulting from the operation of (1) and (2).

Since the quantitative treatment of glacier mass transport, budget, and response to climatic change (Nye, 1960, 1963) is based on the relation between flow rate and stress, and since basal slip contributes a significant to predominant (near the snout) fraction of the total flow, it is important that theoretical concepts of the basal slip process be tested by critical observations and measurements of the actual process in operation. The need for an improved understanding of the phenomenon has been stressed recently by the observation of large anomalous changes in the relative contributions of internal deformation and basal slip over short longitudinal distances in temperate glaciers (Kamb and Shreve, 1963; Savage and Paterson, 1963).

In the study reported below, we have observed directly the basal-slip process in operation, by means of a tunnel driven to the base of Blue Glacier, Mount Olympus, Washington. We have also carried out some simple experiments to provide a comparison with the field observations and with theoretical predictions.

FIELD OBSERVATIONS

Part of the 1962 program of ice-flow investigation of Blue Glacier, Washington, involved excavation of an ice tunnel, which entered the glacier near the top of and along the (true) left margin of the ice fall that separates the lower valley-glacier tongue from the accumulation basins above. For a general description of the glacier see LaChapelle (1959) and C. R. Allen and others (1960). The tunnel, whose plan and profile are shown in Figs. 12.1 and 12.2, reached bedrock 50 m in from the surface and at a vertical depth of 26 m beneath the surface.

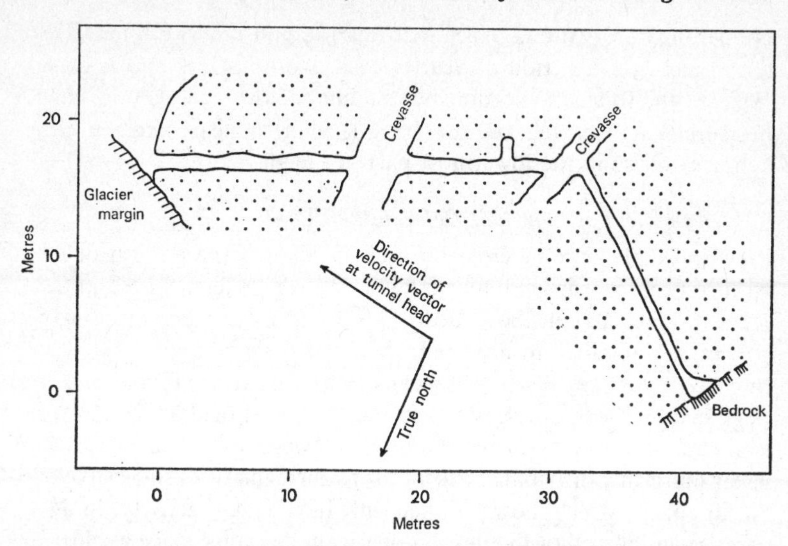

Fig. 12.1 Plan view of tunnel in Blue Glacier, excavated June–July 1962. Observations reported were made in chamber at head of tunnel, adjacent to bedrock

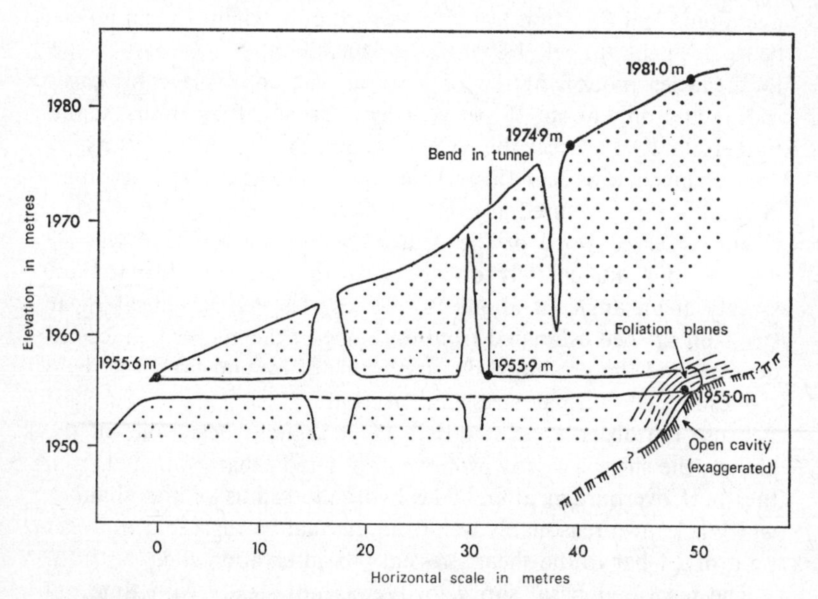

Fig. 12.2 Vertical cross-section along axis of tunnel. Configuration of ice foliation near the tunnel head is indicated, and ice separation from bedrock is shown schematically. Crevasses shown approximately only

Numerous observations of ice deformation and flow were made in the tunnel over a period of nearly 7 weeks during July and August 1962. In the following sections we summarize our observations and measurements bearing on the basal-slip mechanism and on the behavior of ice in the lowermost part of the glacier.

Measurement of Flow and Deformation Rates

Precision dial micrometers were used to observe the motion of ice with respect to bedrock immediately below. A stake anchored in the ice a distance 10 cm above bedrock showed an average motion of 1·6 cm/day parallel to the bed. A stake 150 cm above bedrock moved 12 per cent faster. By means of a vertical profile of marker pegs spaced 10 cm apart, it was found that most of this 12 per cent differential motion occurred within the lowermost 50 cm of ice, as shear uniformly distributed except for two irregularities suggestive of 'shear zones', which, however, had only intermittent activity. In any case, essentially all of the motion of 1·6 cm/day took place as slip at the bedrock–ice interface. Surface flow-velocity measured at a point approximately above the tunnel head was essentially identical in magnitude and direction to that observed at a height 150 cm above bedrock in the tunnel (1·8 cm/day). Thus about 90 per cent of the total glacier motion at the observation site took place by basal sliding and only about 10 per cent by internal deformation. At the tunnel site, located near the glacier margin, the flow velocity was, of course, much less than that in the center of the ice stream some 200 m distant, measured at over 1 m/day.

The observed slope of the bedrock at the tunnel head was 22 degrees, but immediately down-glacier the bedrock dropped off steeply at an angle of about 55 degrees. The average local shear stress on the bed estimated from the slope of the glacier surface (28 degrees) and the presumed density profile of ice and firn is 0·7 bar; the estimated normal stress or overburden pressure is 1·6 bar. Because the observation site is in the marginal shear zone of the glacier, the shear stress is probably larger than that estimated from the local overburden alone. The hydraulic radius of the channel, which is known reasonably well from thermal borings, sets an upper limit of 2·1 bar to the shear stress at the observation site.

The measured basal slip velocity was not steady with time but showed marked irregularities over time intervals of the order of seconds. The mean flow rate varied up to 10 per cent from day to day.

Structural and Textural Observations

When overburden pressure is removed from ice in contact with bedrock, by excavating the ice above and around it, the basal ice freezes fast to the rock. When the basal ice is cut away quickly from the tunnel wall, without initial excavation and release of overburden pressure, it comes free from the sole and is not frozen to it. This observation demonstrates the presence of a thin layer of liquid water, at the pressure melting point, along the ice–bedrock interface. Excavated blocks of basal ice were freed from the sole by irradiating momentarily with a photoflood lamp. The blocks were faced down and polished on the top and sides, to reveal internal structures. After examination in bulk, thin sections cut parallel and perpendicular to the basal slip direction in planes perpendicular to the sole were prepared to reveal the structure and texture of the lowermost 1 m of ice. Seventeen thin sections, made from nine blocks of basal ice, were examined.

Both in block specimen and in thin section, there is revealed a basal ice layer that is structurally and texturally distinct from the ice above. The thickness of this layer as seen in our specimens varies from a maximum of about 2·9 cm to nearly zero where the layer pinches out against bedrock protuberances. On the downstream side of such protuberances the layer reforms. Where it thus reforms, the upper boundary of the newly formed ice layer is always nicely matched to the crest of the preceding obstacle. These features, as well as others detailed below, indicate that the layer is formed by the regelation process. Hence we propose to call it the *regelation layer*.

The regelation layer contains regular trains of fine spherical bubbles or very elongate, tubular cavities aligned in the flow direction. The bubble trains at different levels may mark the heights of different obstacles over which the basal ice has passed in its immediate pre-history. The upper boundary of the regelation layer represents a sharp textural break, as seen under crossed polaroids. The average grain size in the layer is distinctly smaller than in the ice above. Some grains in the regelation layer show crystallographic continuity with grains in the ice above, as would be expected from the point of view of growth nucleation. There is sometimes a suggestion of multiple textural breaks corresponding to the bubble trains at different levels. The upper boundary of the regelation layer always appears perfectly planar as seen in sections cut parallel to the flow direction. In perpendicular sections, it shows some undulation suggestive of transverse topographic irregularities, so that in fact the boundary is a cylindrical surface.

The regelation layer is heavily loaded with debris in comparison with the ice above. The debris content varies markedly from place to place with the layer, but is nowhere greater than about 10 per cent by volume of the layer. The debris consists of fine mud and of rock fragments up to 1 or 2 mm in size. A debris-laden basal layer, which may have been identical to the regelation layer as defined here, was reported by Haefeli (1951) and McCall (1952, p. 128). Rock debris accumulates on the upstream side of bedrock protuberances, from the crest of which there extends downstream a train of debris particles evidently derived from the accumulation on the upstream side.

Where there are depressions below the general local level of the sole, as contrasted with protuberances, the ice does not fill these in but instead bridges over them. This occurs for cavities ranging in width from 4 cm to at least 10 m; the corresponding separation of ice from bedrock ranges from about 1–20 cm. It is difficult from our observations to define precisely the circumstances under which a depression will or will not be filled by the regelation layer, but in general, if a depression is not preceded upstream by a complementary protuberance, it will bc bridged. Where a depression is bridged, there is found forming in the open cavity, just downstream from the point of separation of ice from bedrock, a mass of coarse, needle-like ice particles (length *c.* 5 cm) that we propose to call *regelation spicules*. These spicules are formed in loose aggregates aligned in the direction of ice flow. They are only weakly attached to the ice undersurface, or actually sag or drop away from it. The smaller spicules are single crystals. Although production of the spicules appears to be continuous, the larger open cavities beneath the ice undersurface do not become filled solidly with them, the spicular ice being carried along and presumably reincorporated into the regelation layer at the downstream edge of the open cavities. Spicule ice probably evaporates also, in cases where the open cavities interconnect to the surface so that there can be air flow through them, as observed for larger cavities encountered in the tunnel. The spicule ice, when it is attached firmly enough to the base of the regelation layer that the two can be thin-sectioned together, shows a contrasting texture and in particular a lack of included air bubbles or bubble trains. Spicule formation is inhibited where the ice of the regelation layer is heavily debris-laden. This may be the reason that regelation spicules have not been reported by previous observers (Carol, 1947; Haefeli, 1951; McCall, 1952).

The lowermost 0·5 m of ice above the regelation layer was relatively poor in bubbles and appeared unfoliated. A similar basal zone was found by McCall (1952, p. 128) at depth 50 m in a cirque glacier. Above this zone, typical bubble foliation was prominent. Its attitude was grossly conformable with the sole at the tunnel head, but there was a variation in attitude from one side of the tunnel to the other, outlining a broad anticlinal structure plunging in the direction of flow. It thus appears that the tunnel struck bedrock near the crest of a bedrock prominence. This is further indicated by the configuration of the bedrock observable downstream, below the level of the tunnel floor (Fig. 12.2), where there was ice separation over a distance of about 10 m. The abrupt steepening of the bed is reflected in a bending downward of the ice foliation planes and the regelation layer, as traced down-glacier. This is indicated in Fig. 12.2.

The 'shear zones' suggested by the vertical velocity profile in the basal ice were not reflected in any textural or structural peculiarities in the ice itself, as seen in thin section.

EXPERIMENTS

We have tested the concept of the *regelation layer*, introduced above, by producing a similar feature experimentally. This was done as follows. A 1-cm cube of solid material, to which a constantan wire 0·812 mm in diameter was attached, was frozen into a block of ice about 40 cm in size. A load of 17·5 kg was applied to the constantan wire from the outside, and the ice block allowed to warm gradually to the melting point. The motion of the cube under constant load was measured as a function of time by observing the displacement of the constantan wire against a reference scale anchored in the ice. The arrangement is shown in Fig. 12.3.

The experiment was carried out for cubes of dunite (olivine rock), plexiglass (polymethyl-methacrylate), and aluminum. The observed time–displacement curves are shown in Fig. 12.4. It is inferred that the large acceleration in the motion that took place about 15 hours after the start of the experiments corresponds to the time during warm-up at which the pressure-melting point was reached at the loaded face of the cubes, and hence at which regelation-slip started.

After termination of the experiments, the ice blocks were faced down to thick slabs, examined, and then thin-sectioned through the area of interest around the cubes. The following features were observed:

(1) The ice behind the plexiglass cube, in the volume swept out by its motion, was similar in texture and structure to ice in the regelation

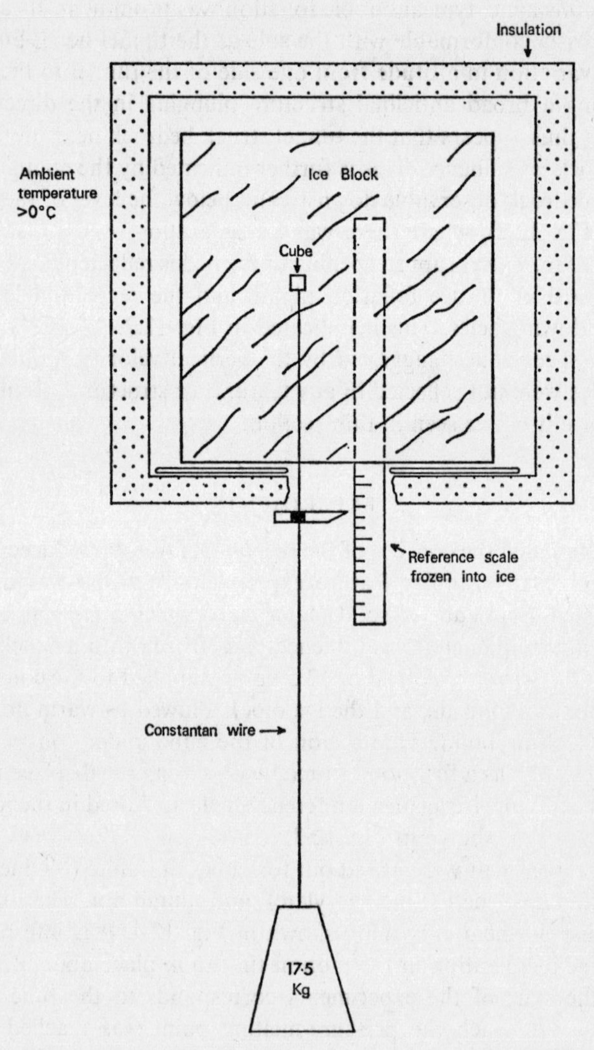

Fig. 12.3 Arrangement for experimental study of regelation flow

layer of the glacier: they have similar grain sizes, similar extent and sharpness of the textural break between regelation zone and adjacent ice, and similar presence of bubble trains parallel to the motion. Refreezing within the regelation zone in this experiment was almost complete.

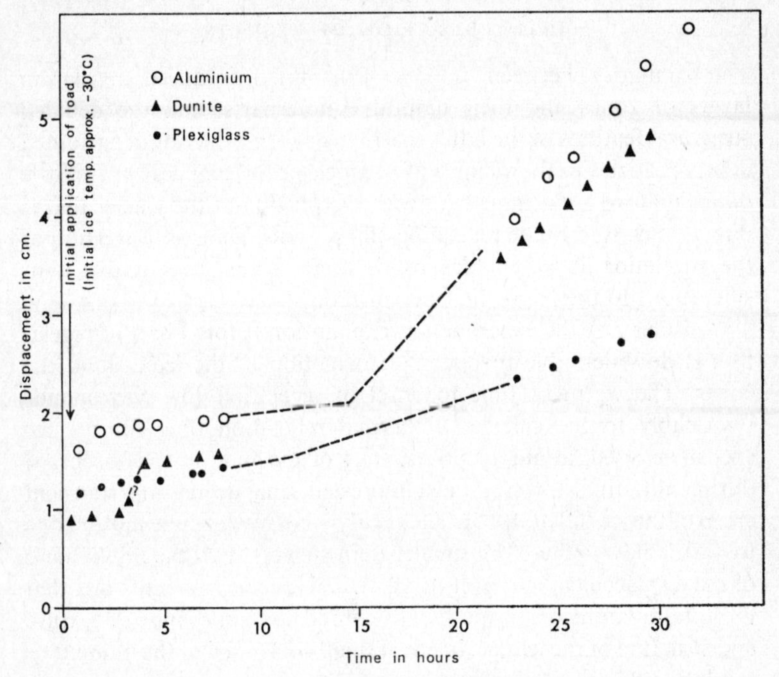

Fig. 12.4 Observed cube displacement as a function of time in the regelation-flow experiments. Observations for the cubes of different materials are indicated by separate symbols as shown. Motion intervening between observed points is indicated hypothetically by dashed lines

(2) The volume swept out by the dunite cube was only partially refrozen, about half remaining as liquid water, containing a small bubble of air and/or water vapor. The walls of the liquid cavity, adjacent to the surrounding host ice, were lined with a regelation layer of ice similar to that in the previous experiment.

(3) Refreezing behind the aluminum cube was also only about 50 per cent effective, but in this case the excess water drained away along the constantan wire, leaving a hollow space behind the cube. Surrounding this there was, again, a regelation layer adjacent to the host ice.

(4) The texture and structure of the host ice around the paths followed by the cubes were not noticeably disturbed, as would have been required if a sizable fraction of the cube motion had been produced by plastic deformation of the ice.

INTERPRETATION OF RESULTS

The similarity between the experimentally produced regelation layers or zones and their natural counterparts, and the detailed structural features of the latter (particularly the behavior of entrained debris particles in the vicinity of bedrock protuberances), leave little doubt that we are dealing here with the portion of the glacier that is directly involved in the regelation-slip process. Direct observation of the regelation layer enables us to apply some quantitative considerations to the basal-slip phenomenon.

Consider first the experiments. The motion before onset of regelation is doubtless due to plastic deformation of the ice around the cubes. The warm-up time to onset of regelation-flow corresponds reasonably to the calculated thermal relaxation time for the ice specimens used, about 10 hours. Part of the motion that occurred during this time interval must represent time-decreasing transient creep of the ice (Glen, 1955). The steady-state rate of cube motion due to plastic flow is therefore smaller than suggested by the initial parts of the displacement curves (Fig. 12.4). This accords with the fact that the total amount of plastic flow recorded in disturbance of texture and structure of the ice specimens is small compared to the amount of regelation flow.

The observed flow rate after onset of regelation may be readily compared with theoretical prediction in the case in which the thermal conductivity of the cube is large compared with that of ice, so that the heat flow is one-dimensional within the cube. In that case

$$v_{\mathrm{R}} = kC\sigma/H\rho L$$

where k is the thermal conductivity of cube material, C the slope of the pressure melting point curve (0·0074°C/bar), $H\rho$ is the heat of ice fusion per unit volume, L the cube edge length, and σ the compressive stress on the loaded face of the cube. The predicted and observed rates are shown in Table 12.1.

Although there is a qualitative parallelism between thermal conductivity and observed flow rate, the quantitative agreement

Table 12.1 Predicted and observed rates of regelation flow

Material	k cal/cm/s/°C	v_R (calculated) cm/day	v (observed) cm/day
Plexiglass	0·012	1·8	1·6
Dunite	0·012	1·8	3·4
Aluminum	0·49	74·0	5·4

between calculated and observed flow rates is not good. The disagreement can probably be accounted for as follows. The loss of meltwater by leakage along the constantan wire, in the case of the aluminum cube, removed to the outside of the ice block part of the heat source needed for regelation flow, so that the flow rate was mainly limited by the conductivity of ice rather than of aluminum. In the case of the dunite cube there was also significant heat conduction from the surface of the ice block, because the refreezing was incomplete. (The work furnished by the applied load is negligible.) The flow rate must therefore have been limited by a higher 'effective conductivity' than that of the dunite alone, otherwise refreezing behind the cube would have been complete. At first sight it seems paradoxical that heat can be conducted from the surface into the interior of an ostensibly isothermal block of ice at the melting point, but in fact the temperature at the stressed face of the cube is below 0°C (actually about −0·1°C) and temperature gradients therefore exist in the ice. It would be necessary to eliminate the effect of this by a different experimental arrangement in order to get quantitative agreement between predicted and observed regelation flow rates; it is significant that the best agreement is obtained in the case where refreezing was essentially complete.

The observations of cube motion are pertinent to the bedrock slip phenomenon in glaciers on account of the fact that Weertman (1957, 1962) based his theory of this phenomenon on the resistance to slip offered by obstacles of cubical shape. If we apply this theory (in its 1962 form) to our slip-rate measurements in Blue Glacier, it appears at first sight to give rather reasonable results. Taking the measured slip rate of about 600 cm/yr and an estimated average shear stress of 0·7 bar at the bed, we compute from Weertman's equation (4) (1962,

p. 32) a 'controlling roughness ratio' $r_c = 9{\cdot}5$ (we used $n = 3{\cdot}2$), and then from equation (3) the 'controlling obstacle size' $L_c = 0{\cdot}56$ cm. The corresponding 'obstacle spacing' is $L'_c = 5{\cdot}3$ cm. Such obstacle sizes and spacings are indeed comparable to those that produce the regelation layer we observed. (For the limiting shear stress of 2·7 bar the theoretical quantities are $r_c = 5{\cdot}5$, $L_c = 0{\cdot}56$ cm, $L'_c = 3{\cdot}1$ cm.)

However, the controlling obstacle size is supposed in the theory to be that for which accommodation of the basal ice to bedrock obstacles is accomplished equally by regelation and by plastic flow, whereas in actual fact there is *no significant plastic deformation* occurring around obstacles of the size and spacing that produce the regelation layer. This is proved in repeated instances by close examination of the regelation layer, whose upper surface is perfectly planar (strictly, cylindrical) and shows no trace of warping even over the largest obstacles that generate the layer. Plastic deformation, indicated by bending of foliation planes and of the regelation layer as traced downstream, becomes evident only over distances of about a metre, much larger than the calculated 'controlling obstacle spacing' of 5·3 cm.

Hence the theoretical 'controlling obstacle size' has no physical meaning in relation to our actual observations. The trouble lies in the predicted distance scales over which regelation or plastic flow should predominate. A similar difficulty with the theory is seen in relation to the cube experiments. If we use the nearly appropriate Fig. 2 of Weertman's 1957 paper, slightly modified to take into account a factor of 2 for the stress value pertinent to plastic deformation under the experimental conditions,* we find that the theory predicts a plastic slip rate that should exceed the regelation slip rate by about an order of magnitude, whereas in actual fact the regelation rate is about an order of magnitude faster, and the amount of plastic deformation, which, if the theory were correct, would have been prominent in the internal structure of the specimens, is small.

Although Weertman's theory thus appears inapplicable quantitatively, we wish to emphasize that we think the general qualitative ideas on which it is based are correct.

* Weertman assumes that the resisting stress on the cube is equally divided between relative compression on the upstream face and relative tension on the downstream face. Since in our experiment there can be no actual tensile stress on the downstream face, liquid water being present, the entire load stress must be applied to the stress to estimate the creep rate, as done by Weertman.

It might be possible by adjusting arbitrarily the constants in Weertman's theory to improve its applicability to our observations, but this does not seem to us a very satisfying approach. Instead, we have developed a new analytical treatment of the basal slip mechanism, which among other things avoids the rather implausible model of cubical obstacles. In our theory, a natural length emerges which is associated primarily with the *spacing* of obstacles and only secondarily, if at all, with their *size*, in contrast to Weertman's approach. This natural length corresponds to the transition from regelation slip to plastic slip, and turns out to have a value of about 0·5–1 m for situations of practical interest, corresponding rather well with our observations of the scale of plastic deformation in the tunnel.

One element of the new theory can be checked directly against our observations of the regelation layer in process of formation. Since we know that the local resistance to basal slip provided by the obstacles that are observed to produce the regelation layer is due solely to the requirements of the regelation process, plastic flow not being involved, we can compute the local average shear stress at the bed from this process alone. We consider a bed roughness having spacings (wavelength of irregularities) $\lambda \approx 4$ cm and 9 cm and sizes (crest to trough amplitudes) $2a \approx 0{\cdot}4$ cm and 1·0 cm respectively. From our theory we can then compute the shear stress τ at the bed from the regelation slip velocity v_{R} and the roughness wavelength λ and amplitude a as follows:

$$\tau = \frac{\pi}{2\sqrt{2}} \frac{H\rho a^2 v_{\mathrm{R}}}{(k_1 + k_2)C\lambda}$$

where $k_1 + k_2$ is the sum of the thermal conductivities of bedrock and ice, taken to be 0·01 cal/cm/s/°C. We obtain 0·2 bar and 0·7 bar respectively. These values are somewhat below the overall shear stress of about 1·5 bar (limiting range 0·7–2·1 bar) estimated mechanically, which is to be expected, since the largest contribution to the shear stress should come from bedrock irregularities on a scale of the natural length mentioned previously.

A complicating factor in the discussion of applicability of the Weertman theory is the relatively low value (1·6 bar) of the normal stress or overburden pressure at the point of observation, which allows a significant amount of ice separation from bedrock. If the Weertman theory were applicable to our observations, so that $L_c = 0{\cdot}5$ cm and $L'_c = 5$ cm were physically meaningful lengths, then bedrock separation should be most prominent over distances of this order, since the 'controlling obstacles' of the theory are those to which the basal ice has the greatest difficulty accommodating (so that the required basal stress concentrations are the largest). For obstacle spacings of the order of metres, much longer than L'_c, plastic flow should allow ready accommodation of the ice to its bed. In actual fact, ice separation is prominent over distances of this order, much more so than over distances of order L'_c. Significant bedrock separations have been observed at overburden pressures up to about 5 bar in other glaciers (Carol, 1947; Haefeli, 1951; McCall, 1952, p. 127). These considerations, as well as our experimental results, all point to the same difficulty with the Weertman theory: that it overemphasizes the contribution to basal-slip of plastic flow as compared with regelation.

The maximum thickness of the regelation layer (1–3 cm in our specimens) is only slightly larger than the crest-to-trough amplitudes of the bedrock irregularities of spacing 5–10 cm that are observed to produce the layer, whereas, according to the interpretation presented here, a thickness of about half the amplitude of the irregularities on a spacing of the order of the natural length mentioned above could be expected to occur. For equal roughness ratios, the expected thicknesses might therefore be almost an order of magnitude larger than those observed. Several factors probably contribute to the effect: (1) The observation site is near the crest of a bedrock prominence, where the regelation layer corresponding to bedrock topography wavelengths of the order of a metre or longer would have minimum thickness; in conformity with this, the thickest part of the regelation layer (3 cm) is found in the specimen of ice examined from farthest downstream. (2) In the bedrock area actually observed, there seems by chance to be a scarcity of well-defined irregularities of wavelength near 50 cm. (3) The tendency for bedrock separation to occur prominently over distances of the order of 1 m or more effectively reduces the amplitude of irregularities of these wavelengths. (4) It is possible that thicker regelation layers inherited from farther upstream could have been obliterated by recrystallization, or by progressive melting without concomitant refreezing at the sole. The question of the 'ultimate' thickness of the regelation layer can be discussed more thoroughly on a theoretical basis. In the present paper we wish to point out only that the natural distance scale for transition from regelation slip to plastic slip expresses itself basically in terms of the wavelength of the irregularities rather than in terms of their amplitude. The thickness of the regelation layer, which involves the effective bed roughness, is of less basic significance.

REFERENCES

ALLEN, C. R., *et al.* (1960) 'Structure of the lower Blue Glacier, Washington', by C. R. Allen, W. B. Kamb, M. F. Meier, and R. P. Sharp, *Journal of Geology*, LXVIII 6, pp. 601–25.

FARADAY, M. (1860) 'Note on regelation', *Proceedings of the Royal Society*, X 440–50.

GERRARD, J. A. F., *et al.* (1952) 'Measurement of the velocity distribution along a vertical line through a glacier', by J. A. F. Gerrard, M. F. Perutz, and A. Roch, *Proceedings of the Royal Society*, ser. A, CCXIII 1115, pp. 546–58.

GLEN, J. W. (1955) 'The creep of polycrystalline ice', *Proceedings of the Royal Society*, ser. A, CCXXVIII 1175, pp. 519–38.

GLEN, J. W. (1961) 'Measurement of the strain of a glacier snout', *Union Géodésique et Géophysique Internationale. Association Internationale d'Hydrologie Scientifique. Helsinki Conference, 1960. Commission des Neiges et Glaces*, pp. 562–7.

HAEFELI, R. (1951) 'Some observations on glacier flow', *Journal of Glaciology*, I 9, pp. 496–500.

KAMB, W. B., and SHREVE, R. L. (1963) 'Structure of ice at depth in a temperate glacier', *Transactions American Geophysical Union*, XLIV 1, p. 103. [Abstract].

KINGERY, W. D. (1960) 'Regelation, surface diffusion, and ice sintering', *Journal of Applied Physics*, XXXI 5, pp. 833–8.

LACHAPELLE, E. R. (1959) 'Annual mass and energy exchange on the Blue Glacier', *Journal of Geophysical Research*, LXIV 4, pp. 443–9.

MATHEWS, W. H. (1959) 'Vertical distribution of velocity in Salmon Glacier, British Columbia', *Journal of Glaciology*, III 26, pp. 448–54.

NYE, J. F. (1960) 'The response of glaciers and ice-sheets to seasonal and climatic changes'. *Proceedings of the Royal Society*, ser. A, CCLVI 1287, pp. 559–84.

NYE, J. F. (1963) 'Theory of glacier variations', in W. D. Kingery, (ed.), *Ice and Snow; Properties, Processes, and Applications: Proceedings of a Conference held at the Massachusetts Institute of Technology, 1962* (Cambridge, Mass.: M.I.T. Press) 151–61.

SAVAGE, J. C., and PATERSON, W. S. B. (1963) 'Borehole measurements in the Athabasca Glacier', *Journal of Geophysical Research*, LXVIII 15, pp. 4521–36.

SHARP, R. P. (1954) 'Glacier flow: a review', *Bulletin of the Geological Society of America*, LVX 9, pp. 821–38.

SHREVE, R. L. (1961) 'The borehole experiment on Blue Glacier, Washington', *Union Géodésique et Géophysique Internationale. Association Internationale d'Hydrologie Scientifique. Helsinki Conference, 1960. Commission des Neiges et Glaces*, pp. 530–1.

THOMPSON, J. (1860) 'On recent theories and experiments regarding ice at or near its melting point', *Proceedings of the Royal Society*, X 152–60.

WEERTMAN, J. (1962) 'Catastrophic glacier advances', *Union Géodésique et Géophysique Internationale. Association Internationale d'Hydrologie Scientifique. Obergurgl Colloquium 1962. Commission des Neiges et Glaces*, pp. 31–9.

13 The Theory of Glacier Sliding

J. WEERTMAN

UNTIL recently the amount of experimental research devoted to the study of the sliding of glaciers has been quite limited. Whatever understanding we had of this phenomenon came principally from theoretical work (Weertman, 1957, 1958, 1962; L. Lliboutry, 1959, 1963). The situation now is changing. B. Kamb and E. LaChapelle (1963, 1964) have carried out extremely interesting field studies and laboratory tests on the mechanisms involved in glacier sliding. Lliboutry and R. Brepson (1963) have constructed a large machine in which 30 kg blocks of ice will be made to slide. G. R. Elliston (1963) has shown from field work on the Gornergletscher that (melt) water at the bottom of a glacier profoundly influences the sliding velocity.

It seems likely that the phenomenon of glacier sliding will be the subject of an active field of research in the near future. Obviously data have been and will be obtained which can be used to test quantitatively the theories on glacier sliding. It is desirable that the theories themselves be developed as completely as possible for these tests.

It is the purpose of this paper to develop a sliding theory which is more general than that previously presented (Weertman, 1957, 1958, 1962). One improvement incorporated into the new theory is the fact that, whereas in the previous version all resistance to sliding comes from a 'controlling protuberance size', now the resistance produced by other sizes of obstacles is considered. The resistance offered by the other obstacles is smaller than that caused by the controlling obstacles; nevertheless it is appreciable and should be taken into account.

Another improvement in the present theory comes from the relaxation of the assumption made in the older version that the hydrostatic pressure at the bottom of a glacier always is larger than any possible tensile stress occurring there. In the newer theory account is taken of situations in which this assumption is not valid.

THEORY

Two sliding mechanisms form the basis of the theory. One of these involves the phenomenon of pressure melting. In this mechanism ice is melted on the upstream, high-pressure side of an obstacle. The water produced flows around the obstacle to the low-pressure side where it refreezes. The velocity of melting and freezing, and thus of ice motion, is determined by the temperature gradient across the obstacle. This gradient is larger the smaller the obstacle, and thus the speed of sliding is larger the smaller the obstacle size. The pressure-melting mechanism permits relatively fast ice motion past small obstacles but not around large obstacles. A second sliding mechanism was introduced in order to obtain motion of ice around large protuberances. This mechanism is based on the enhancement of the creep rate caused by stress concentrations existing near obstacles. It leads to a sliding velocity which increases with increasing obstacle size. The existence of both of these mechanisms has been verified by the field observations of Kamb and LaChapelle at the end of the ice tunnel in Blue Glacier. The basis of our theory is thus established and no longer need be regarded as speculative.

In my original paper (1957) I postulated the existence of an idealized glacier bed containing cubical obstacles. The assumption that the obstacles have a cubic shape was one of convenience. It is obvious that essentially identical results would be obtained from the analysis if the exact shape of the obstacles were left unspecified and only their average dimension were used in the equations. In order to make the sliding theory more general we shall consider in this paper obstacles whose three dimensions do differ from one another. We let L_h represent the average height of an obstacle, and L_d and L_p represent respectively the average widths in the direction of glacier flow and in the direction perpendicular to the flow. It is not necessary to specify the exact shape of the obstacles.

It was shown in the original paper that, if the obstacles in a glacier bed are all of the same size, a definite sliding velocity can be calculated from each of the two sliding mechanisms. In order to make this calculation it is necessary to assume that a shear stress cannot be supported across a smooth rock–ice interface. This assumption obviously is valid if the ice is at the melting point and a thin film of water exists between the rock and the ice. It is not valid if the ice is below its melting point. Thus cold glaciers or ice sheets should not slip at their bed, a conclusion borne out by at least one field observation (R. P. Goldthwait, 1960).

Consider a bed containing obstacles of uniform size which are separated from one another by an average distance L'. The average force exerted on any one obstacle is $\tau L'^2$ when a shear stress τ acts parallel to the bed, provided that a film of water exists between the rock at the bed, and the ice of the glacier. If the ice exerts a force on an obstacle then conversely the obstacle pushes through the ice with the same force $\tau L'^2$. Since the average cross-sectional area of an obstacle is $L_h L_p$, this force produces a compressional stress approximately equal to $\tau L'^2/L_h L_p$ on the upstream side of the obstacle. A stress of this magnitude should exist within a volume of ice of the same size as the obstacle itself.

In his review of the original (unpublished) version of my first paper (Weertman, 1957) on glacier sliding J. W. Glen, in a private communication, pointed out that the force exerted on the ice by an obstacle results not only in the existence of a compressive stress on the upstream side of the obstacle but it also may cause the ice to experience a tensile stress on the downstream side. In order that this tension exist it is necessary that the ice does not lose contact with the rock surface. Thus if the hydrostatic pressure is great enough to prevent a cavity from being formed on the downstream side of an obstacle, the obstacle produces not only compression in the ice on its upstream side but also tension on its downstream side. (The tensile and compressive stresses we are discussing are stresses additional to the hydrostatic pressure normally present at the bottom of the glacier.) It is obvious that if the obstacle is symmetrical the tensile stress is of the same magnitude as the compressive stress. If ice is to close in behind an obstacle as it flows around it the flow lines on the upstream and downstream side of a symmetrical obstacle must be symmetrical and the stresses causing this flow likewise must be symmetrical. Therefore when Glen's condition is valid and the compressive stress is $\tau L'^2/2L_h L_p$, the tensile stress also is equal to $\tau L'^2/2L_h L_p$. The combination of these two stresses represents a total force of $\tau L'^2$ exerted on the obstacle.

If the hydrostatic pressure is not great enough to prevent the formation of a cavity on the downstream side of an obstacle the compressive stress on the upstream side is $\tau L'^2/L_h L_p$, a value which is twice as great as that which is realized when Glen's condition holds.*

* If the ice is cold and frozen to a rock surface the ice–rock interface can support a tensile force normal to it. In this situation a tensile stress could exist on the downstream side of an obstacle even when the hydrostatic pressure is small. It also should be realized that, even if ice is at the melting point, an ice–water–rock interface might be able to support a tensile force across it since liquids do have an appreciable tensile strength. The hydrostatic tensile strength of water

In the published version of the sliding theory (Weertman, 1957) it was assumed that Glen's condition always is satisfied.

In this paper we shall consider situations in which Glen's condition is valid and those in which it is not.

PRESSURE MELTING

The velocity of sliding caused by pressure melting is found from a calculation of the change in hydrostatic pressure from one side of an obstacle to another. Because the melting temperature of ice varies with pressure, there is a temperature difference across the obstacle which gives rise to the flow of heat, the melting of ice, and the freezing of water. The difference in temperature ΔT is equal to $C\Delta P$, where C is a constant equal to $7{\cdot}4 \times 10^{-9}$ °C/cm²/dyn. and ΔP is the difference in hydrostatic pressure.

There is some ambiguity connected with the pressure difference ΔP. In the water layer existing between the ice and rock the pressure difference will be $\Delta P = \tau L'^2/L_h L_p$. In the ice itself, however, the difference may be only $\tau L'^2/3L_h L_p$. The factor of $\frac{1}{3}$ in this latter expression arises because a uniaxial compression or tension produces a hydrostatic pressure of only $\frac{1}{3}$ the magnitude of the compressive or tensile stress. (That is, through a rotation of the coordinate system, a uniaxial stress in one coordinate system can be changed in another coordinate system to a stress system containing only pure shear stresses and a hydrostatic stress. The value of the hydrostatic stress turns out to be $\frac{1}{3}$ the value of the original uniaxial compressive or tensile stress.) In our previous papers on sliding we assumed that the factor $\frac{1}{3}$ should appear in the equations. On the other hand, Kamb

---

has been measured by Briggs (1950). He finds it to rise steeply with temperature from 20 bars at a temperature slightly above 0°C to 280 bars at temperatures between 5° to 10°C. J. C. Fisher (1950) has pointed out that the drop in strength by an order of magnitude with a small decrease in temperature may be caused by the nucleation of ice under the reduced hydrostatic pressure. It would be desirable to have an experimental measure of the tensile strength of water in contact with an ice surface to see if it still has a finite value under these conditions.

For a liquid to have a tensile strength it must be in a confined space. This condition may not be met at the bottom of a glacier and the tensile strength of the water layer could be zero. Nevertheless in laboratory experiments, such as those carried out by Kamb and LaChapelle in which blocks of rock are pulled through ice, it is possible that a tensile stress is exerted across the water layer separating ice from rock since in this type of experiment the water layer may be confined.

and LaChapelle (1964) consider that it should not. This is a question best answered by experiment, and Kamb and LaChapelle's (limited) experimental data do seem to support their viewpoint. Therefore in what follows we shall consider that the temperature difference across an obstacle is $C_T L'^2/L_h L_p$, rather than the smaller value we used previously. The temperature gradient across the obstacle is $C_T L'^2/L_h L_p L_d$. If all the heat flow is through the obstacle a volume of ice is melted in a unit time which equals the temperature gradient $(C_T L'^2/L_h L_p L_d)$ times the area of an obstacle $(L_h L_p)$ times the coefficient of thermal conductivity D of the rock divided by $H\rho$, where H is the heat of fusion of ice (80 cal/g) and ρ is the density of ice. The velocity of sliding is equal to this volume of melted ice divided by the area of an obstacle. The following equation is obtained for the velocity of the sliding S_1 which results from pressure melting:

$$S_1 = C_T L'^2 D/H\rho L_h L_p L_d = (C_T D/H\rho L)(L'^2/L^2) \tag{1a}$$

where L is the average dimension of an obstacle $(L^3 = L_h L_p L_d)$. Except for the absence of a factor of $\frac{1}{3}$, equation (1a) is identical to the equation previously obtained for the pressure-melting mechanism. Equation (1a) is derived with the assumption that all the heat flows through the obstacle and none goes through the surrounding ice. To take this latter heat flow into account we rewrite equation (1a) as follows:

$$S_1 = aC_T L'^2 D/H\rho L_h L_p L_d = (aC_T D/H\rho L)(L'^2/L^2) \tag{1b}$$

where a is a constant whose value is 1 if all of the heat flow is confined to the obstacle and is somewhat larger than 1 if additional heat flows through the ice. The value of a could be determined by laboratory experiments.

STRESS CONCENTRATIONS

The velocity of the sliding which results from the creep-rate enhancement caused by stress concentrations is found by noting that the volume of ice which is subjected to the concentrated stress is of the order of the volume of the obstacle itself. The creep rate $\dot{\varepsilon}$ of ice is given by Glen's creep law. It is

$$\dot{\varepsilon} = B\sigma^n \tag{2}$$

where σ is the stress, n is a constant equal to 3 or 4, and B is another constant which is equal to 0·017 bar^{-n}/yr when the stress is uniaxial tensile or compressive. The compressive stress σ on the upstream side of an obstacle is equal to $\tau L'^2/\beta L_h L_p$, where $\beta = 2$ if the hydrostatic pressure is sufficiently large to prevent cavity formation and $\beta = 1$ if it is not. The sliding velocity S_2 caused by the stress concentration is equal to the creep rate $\dot{\varepsilon}$ times the distance in the direction of motion over which this creep rate is effective. Thus

$$S_2 = \dot{\varepsilon} L_d = BL_d(\tau L_d/\beta L)^n(L'^2/L^2)^n \tag{3a}$$

where, as before, L is the average dimension of an obstacle. This sliding velocity is almost the same as that derived previously. The sliding velocity given by equation (3a) is, of course, only a rough estimate of the sliding velocity S_2 since we do not know exactly the distance over which the creep rate $\dot{\varepsilon}$ is effective. A more exact expression for the sliding velocity is

$$S_2 = b\dot{\varepsilon} L_d = bBL_d(\tau L_d/\beta L)^n(L'^2/L^2)^n \tag{3b}$$

where b is a constant of the order of 1. The exact value of b could be determined from laboratory experiments.

DOUBLE-VALUED NATURE OF THE SLIDING VELOCITY DUE TO STRESS CONCENTRATIONS

It should be emphasized that, in a certain range of values for the overburden pressure, the sliding velocity S_2 actually is a double-valued function of the shear stress τ. The reason for this multiplicity of value is the fact that β may take on one of two possible values. We have noted that if the hydrostatic pressure is very large $\beta = 2$ but if it is small $\beta = 1$. Suppose, however, that the hydrostatic pressure has an intermediate value. Let P equal the hydrostatic pressure, where $P = \rho g h$, ρ is the density of ice, g is the gravitational acceleration, and h is the thickness of the glacier. Let T represent the tensile stress on the downstream side of an obstacle. This tensile stress is exerted in a direction parallel to the bed. Further, let θ represent the maximum angle between the slope of the obstacle and the average slope of the bed. Thus θ is 90° for an obstacle having part of its surface perpendicular to the average slope of the bed. The maximum tensile stress exerted normal to an ice–rock surface is $T \sin^2 \theta$. In order for a tensile stress to exist on the downstream side of an obstacle the ice must remain in contact with the rock and thus the hydrostatic pressure must be larger than $T \sin^2 \theta$, where $T = \tau L'^2/2L_h L_p$.

If the ice loses contact with the downstream surface of an obstacle only a compressive stress exists on the upstream side of the obstacle. Its magnitude is $\tau L'^2/L_h L_p$. A cavity will form on the downstream side. The hydrostatic pressure will tend to close the cavity. The rate of closure (J. F. Nye, 1953) is proportional to P^n. However, the cavity is being opened up by ice flow around the obstacle. If P is greater than $\tau L'^2/L_h L_p$, a cavity will not be able to form. On the other hand, if P is less than $\tau L'^2/L_h L_p$ a cavity can remain open. The value of P can be smaller than $\tau L'^2/L_h L_p$ and yet be larger than $TL'^2 \sin^2 \theta/2L_h L_p$, the stress which it must exceed to prevent an ice–rock interface surface from separating and a cavity being formed. Therefore, at any value of $P = \rho g h$ in the range

$$\tau L'^2 \sin^2 \theta/2L_h L_p < \rho g h < \tau L'^2/L_h L_p$$

a cavity which is formed behind an obstacle will remain open but, if the cavity is not already in existence, it will not form. Therefore β can equal either 1 or 2 in this range of values of P. As a result the sliding velocity is double valued.

THE GLACIER SLIDING VELOCITY

In our previous papers on sliding it was found that the actual sliding velocity of a glacier whose bed contains a full spectrum of obstacle sizes is determined by those obstacles for which $S_1 = S_2$. It was argued that the pressure-melting mechanism enables ice to flow easily around smaller obstacles because this mechanism gives a sliding velocity which is larger the smaller the obstacle. It was argued further that the stress-concentration mechanism enables ice to flow around the larger obstacles without any trouble since for this mechanism the larger the obstacle the greater is the sliding velocity.

We still hold to this viewpoint but we wish to refine the calculation of the actual sliding velocity. In the previous paper it was assumed implicitly that the 'controlling' obstacles, that is, those obstacles whose size is such that S_1 and S_2 of equations (1b) and (3) are equal, hinder the ice flow. The resistance offered by obstacles both smaller and larger than the controlling obstacles was considered to be negligibly small. This assumption obviously is only an approximation. The smaller and the larger obstacles also hinder the ice motion although not to the same degree as the controlling protuberances.

We wish now to calculate the sliding velocity of a glacier when the effect of obstacles both smaller and larger than the controlling protuberances is taken into account.

It is postulated that the glacier bed is made up of obstacles of various sizes. One simplification will be made in the distribution of obstacle sizes. It will be assumed that instead of being continuous the spectrum of sizes is discrete. If λ is regarded as the average dimension of the smallest obstacle the next largest obstacle will be taken to have the average dimension 10λ, the next largest 100λ, the next 1000λ and so on to the largest sized obstacles (which could be of the order of 1/10 of the thickness of the glacier).

Because of the existence of the shear stress τ, a glacier transmits over an area A a total force τA to the bed of the glacier. This force is transmitted through the obstacles and thus each size group will transmit some fraction of the total force. The obstacles in the controlling size group will, of course, transmit the major portion of the force. Suppose we let $\tau_i A$ represent that force which the obstacles of size $(10)^i\lambda$ transmit to the bed. We must have the condition that $\Sigma\tau_i A = \tau A$, or

$$\sum_{i=0} \tau_i = \tau. \tag{5}$$

The stress τ_i can be regarded as the effective shear stress which causes the flow of ice around obstacles of the size $(10)^i\lambda$. Since the sliding velocity must be the same for every sized obstacle it follows that

$$\begin{aligned} S &= (S_1 + S_2)_i \\ &= (aC\tau_i D/H\rho L_i)(L_i'^2/L_i^2) \\ &\quad + bBL_{a_i}(\tau_i L_{a_i}/\beta_i L_i)^n (L_i^2/L_i^2)^n \end{aligned} \tag{6}$$

where S is the actual sliding velocity of the glacier and the subscripts refer to the values of L, L', etc., for the particular size of obstacle.

Equation (6) gives τ_i in terms of the sliding velocity S. If all of the values of τ_i are substituted into the summation term of equation (5) an equation is obtained which contains only the sliding velocity S and the applied shear stress τ as the independent variables. The sliding velocity has been determined as a function of the shear stress acting at the bed of a glacier. By inspection it is obvious that this equation is a complicated one. A good approximation to this more exact equation can be found by observing that in equation (6) the first term on the right-hand side is predominant for obstacles smaller than the controlling size whereas the last term predominates for obstacles larger than the controlling size. Thus we can make the approximation that $S = (S_1)_i$ for the smaller obstacles, that $S = (S_2)_i$ for the larger, and that $S = (S_1 + S_2) = 2S_1 = 2S_2$ for the controlling obstacle size. (Hereafter the controlling obstacle size will be defined to be the size of that obstacle for which the two sliding mechanisms give identical contributions to the sliding velocity.)

Now let Λ represent the average overall dimension and Λ_d the dimension in the direction of motion of the controlling obstacles. Let us assume that the 'roughness' of the bed is not a function of the size of obstacles. Therefore $L'_i/L_i = r$ is constant for all obstacle sizes. The term r is a measure of the roughness of the bed. Also let $L_{d_i}/L_i = \gamma$ be a constant independent of obstacle size. Let τ_Λ represent the effective shear stress acting on the controlling obstacles. Since the sliding velocity is identical for all obstacles the effective shear stresses acting on obstacles smaller than the controlling size are $\tau_\Lambda/5$, $\tau_\Lambda/50$, $\tau_\Lambda/500$, etc. Similarly, the effective shear stresses acting on the larger obstacles are $\tau_\Lambda(\beta/\beta_\Lambda)/5^{1/n}$, $\tau_\Lambda(\beta/\beta_\Lambda)/50^{1/n}$, $\tau_\Lambda(\beta/\beta_\Lambda)/500^{1/n}$ etc., where β_Λ is the value of β for the controlling obstacles.

With these values for the stresses τ_i equation (5) reduces to

$$\tau = \tau_\Lambda[(11/9) + 2^{1/n}/\{(10)^{1/n} - 1\}] \tag{7a}$$

when the value of β for all the larger obstacles is identical to β_Λ of the controlling obstacles. Since $n \approx 3$ this last equation reduces to

$$\tau = 2{\cdot}314\tau_\Lambda. \tag{7b}$$

If $\beta_\Lambda = 1$ for the controlling obstacle size but β of the larger obstacles is equal to 2, equation (7b) becomes

$$\tau = 3{\cdot}405\tau_\Lambda. \tag{7c}$$

Equations (7b) and (7c) represent two limiting cases.

It can be seen that the effective shear stress acting on the controlling protuberances is only about $\frac{1}{2}$ to $\frac{1}{3}$ of the applied shear stress. Thus an appreciable part of the resistance to sliding comes from obstacles other than the controlling obstacles. This additional resistance is provided principally by the obstacles which are larger than the controlling size.

If we substitute equations (7) into equation (6) and set $L_i = \Lambda$ and $S_1 = S_2$, we obtain the following expression for the sliding velocity of a glacier

$$S = 2aCDr^2\tau/kH\rho\Lambda \tag{8a}$$

where k represents the constant term of equations (7) and has a value which lies between 2·314 and 3·405. An expression for Λ is derived from the condition that $S_1 = S_2$ at the controlling obstacle size:

$$\Lambda = (aCDk^{n-1}\beta_\Lambda{}^n/H\rho bBr^{2n-2}\gamma^{n-1}\tau^{n-1})^{\frac{1}{2}}. \tag{9}$$

The substitution of this expression into (8a) results in the equation

$$S = 2(aCDbB\gamma^{n-1}/H\rho\beta_\Lambda{}^n)^{\frac{1}{2}}(\tau r^2/k)^{(n+1)/2}. \tag{8b}$$

This equation gives the sliding velocity of a glacier as a function of the shear stress acting at the bed of a glacier. Apart from a constant factor it is identical to the sliding velocity calculated previously. Both the sliding velocity and the controlling obstacle size derived previously (Weertman, 1957, 1962) can be obtained from these last two equations by setting $k = b = \gamma = 1$, $a = \frac{1}{3}$ and $\beta_\Lambda = 2$, and by dividing the right-hand side of equation (8b) by 2.

Fig. 13.1 shows plots of sliding velocity at constant stress versus the roughness factor r of the bed. (The larger is r the smoother is the bed.) Also shown in these plots is Λ, the controlling obstacle size, versus r. In the calculations for these plots it was assumed that $a = 1$, $n = 3$, $\gamma = 1$, $b = 1$, $C = 7{\cdot}4 \times 10^{-3}$ °C/bar, $B = 0{\cdot}017$ bar^{-3}/yr, $D = 0{\cdot}005$ cal/°C/s (a typical value for rocks), and $\tau = 1{\cdot}0$ bar. In the figure one set of curves was found for $\beta_\Lambda = 1$ and $k = 2{\cdot}31$ and another set for $\beta_\Lambda = 2$ and $k = 2{\cdot}31$. These values of β_Λ and k represent limiting cases. The set of values $\beta_\Lambda = 1$ and $k = 3{\cdot}405$ correspond to curves which are intermediate to those plotted.

In these plots the controlling obstacle size at a particular sliding velocity can be found by drawing a vertical line from the sliding velocity curve to the obstacle size curve. This is done in Fig. 13.1 for a sliding velocity of 80 m/yr, which is of the order of the sliding velocity of glaciers having a shear stress around 1 bar at their beds. The roughness factor of a glacier which slides with the velocity of 80 m/yr under a stress of 1 bar must be of the order of $r = 15$–20. Also shown in Fig. 13.1 are the obstacle sizes and roughness factors

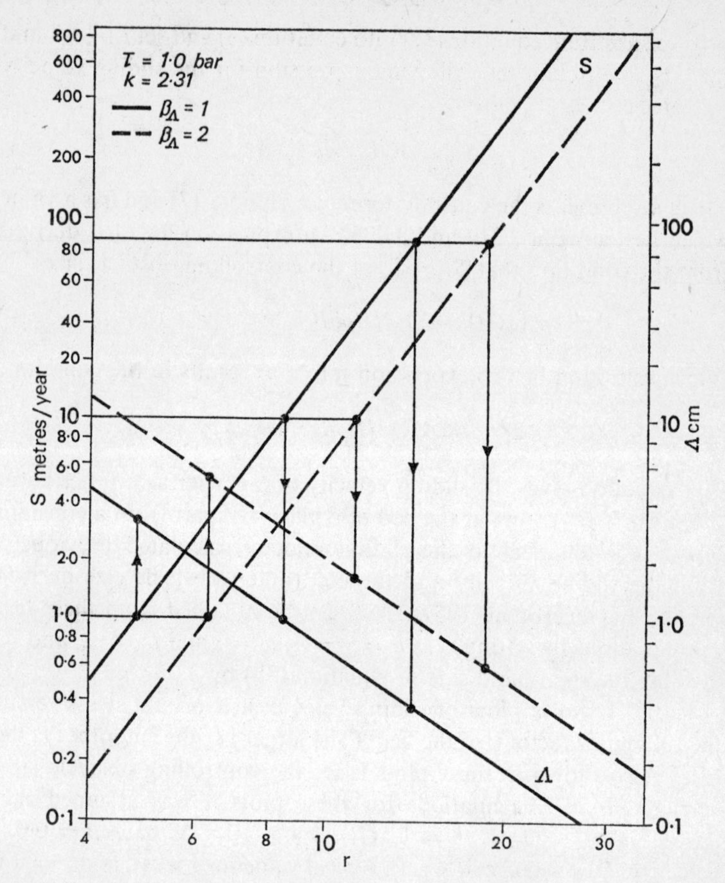

Fig. 13.1 Plot of sliding velocity S and size of controlling obstacles Λ as a function of the roughness factor r for a shear stress $\tau = 1$ bar, $k = 2{\cdot}31$ and $\beta_\Lambda = 1$ or 2. (The case of $k = 3{\cdot}4$ and $\beta_\Lambda = 1$ falls between the curves shown)

connected with sliding velocities of 10 m/yr and 1·0 m/yr under a 1 bar shear stress. A somewhat different method of presenting the results of equations (8) and (9) is shown in Figs. 13.1 and 13.2. Here we find plotted the sliding velocity and controlling obstacle size as a function of the shear stress for various values of the bed roughness.

Kamb and LaChapelle (1964) measured a sliding rate of 5·8 m/yr at the end of their ice tunnel. They estimated that the roughness factor of the bed was equal to 9 and the shear stress acting on the bed was 0·7 bar. For these values of the roughness and the shear stress

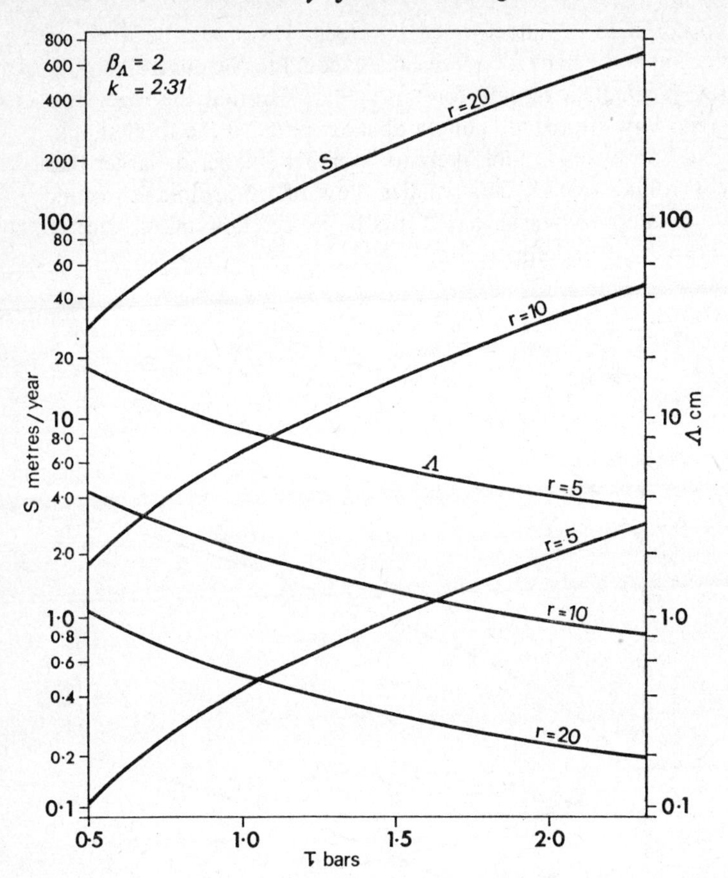

Fig. 13.2 Plot of sliding velocity S and controlling obstacle size Λ as a function of the shear stress τ for various values of the roughness factor r and k = 2·31 and $\beta_\Lambda = 2$

Fig. 13.2 predicts a sliding rate of 2 m/yr, a value close to the observed rate. The controlling obstacle size obtained from this plot is about 4 cm. This size is comparable to the maximum thickness (about 3 cm) of the observed regelation layer. If the maximum thickness of the regelation layer is taken as an approximate measure of the controlling obstacle size, it would appear that theory and observation agree reasonably well. Nevertheless, Kamb and LaChapelle feel that the controlling obstacle size is much greater because the regelation layer behind the obstacles they examined appeared to be undeformed by creep. However, the likelihood of finding an obstacle of a size exactly equal to the controlling obstacle size is small. It is only for this special size that the regelation and creep flow contribute comparable amounts to the total slippage. A given obstacle is more likely to be either smaller or larger than the controlling size. If it is smaller, flow of ice around it occurs predominately by regelation; if it is larger the flow takes place largely through creep deformation.

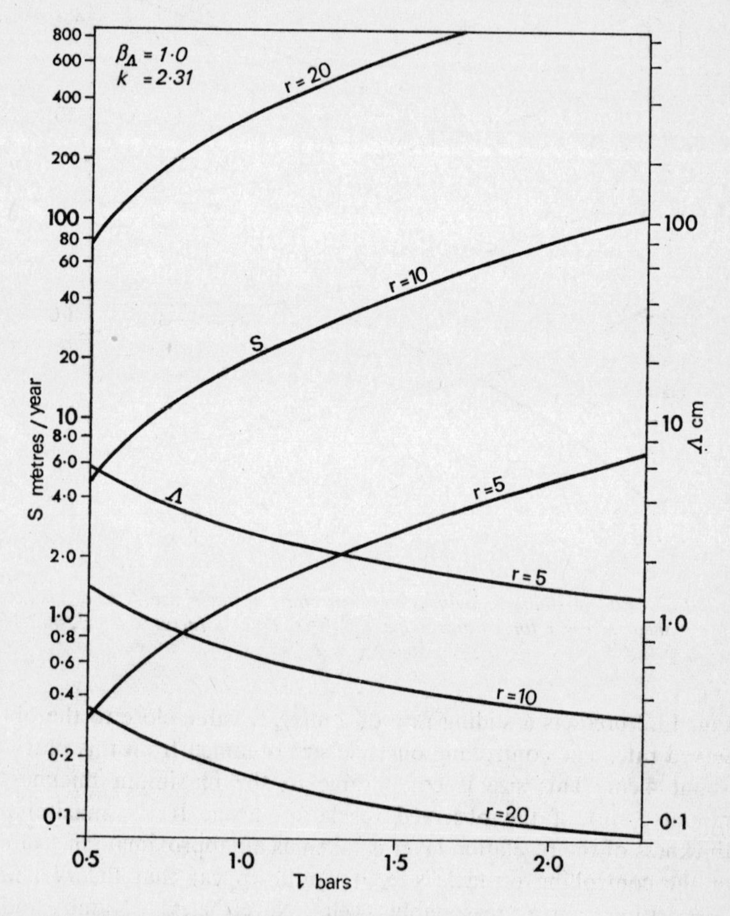

Fig. 13.3 Plot of sliding velocity S and controlling obstacle size Λ as a function of the shear stress τ for various values of the roughness factor r and k = 2·31 and $\beta_\Lambda = 1$

GLEN'S CONDITION

The values of the sliding velocity and the controlling obstacle size depend upon whether or not Glen's condition is satisfied. Glen's condition is satisfied if the hydrostatic pressure ρgh is sufficiently large.

The effective stress on the controlling obstacles is $\tau r^2/\beta_\Lambda k$, where $\beta_\Lambda = 2$ when Glen's condition holds and $\beta_\Lambda = 1$ when it does not. If the hydrostatic pressure ρgh is greater than $\tau r^2/k$, Glen's condition always is valid; if ρgh is less then $\tau r^2 \sin^2 \theta/2k$, Glen's condition is never valid (θ is the angle between the maximum slope of an obstacle and the average slope of the bed); but in the region where

$$\tau r^2 \sin^2 \theta/2k < \rho gh < \tau r^2/k$$

Glen's condition may or may not be valid. In this region the sliding velocity is double-valued. (It is interesting to note that M. F. Meier (private communication) has some field observations which indicate, through an indirect calculation, the occurrence of a double value in the sliding velocity of Nisqually Glacier.)

Figs. 13.4 and 13.5 show regions of validity and non-validity of Glen's condition. In these figures, glacier thickness h is plotted against shear stress for various values of the roughness factor r. The intermediate areas of these plots indicate the region within which β_Λ can have either one of two values. Here the sliding velocity is double valued.

When Glen's condition does not hold, cavities form behind the controlling obstacles and also possibly behind obstacles larger than those of the controlling size. Lliboutry (1959) first predicted the existence of cavities at the bed of a glacier. His prediction was based on a model of a glacier bed, his washboard model, which is rather different from the one we have employed. Nevertheless the physical reason for the appearance of cavities in our model of a glacier bed is much the same as in his.

The length of the cavity which is formed behind an obstacle can be estimated as follows. The stress which causes sliding around an obstacle is $\tau r^2/k$, where $\beta = 1$. The sliding velocity is proportional to $(\tau r^2/k)^n$. Since the overburden pressure is ρgh the velocity with which the cavity is closed is proportional to $(\rho gh)^n$. Therefore the cavity behind an obstacle of size Λ will be closed off at a distance from the obstacle which is of the order of $\Lambda(\tau r^2/k)^n/(\rho gh)^n$. The cross-sectional area of the cavity approaches zero at this distance. At the head of the cavity the cross-sectional area is that of the obstacle itself. The probability that another obstacle of size Λ is directly downstream

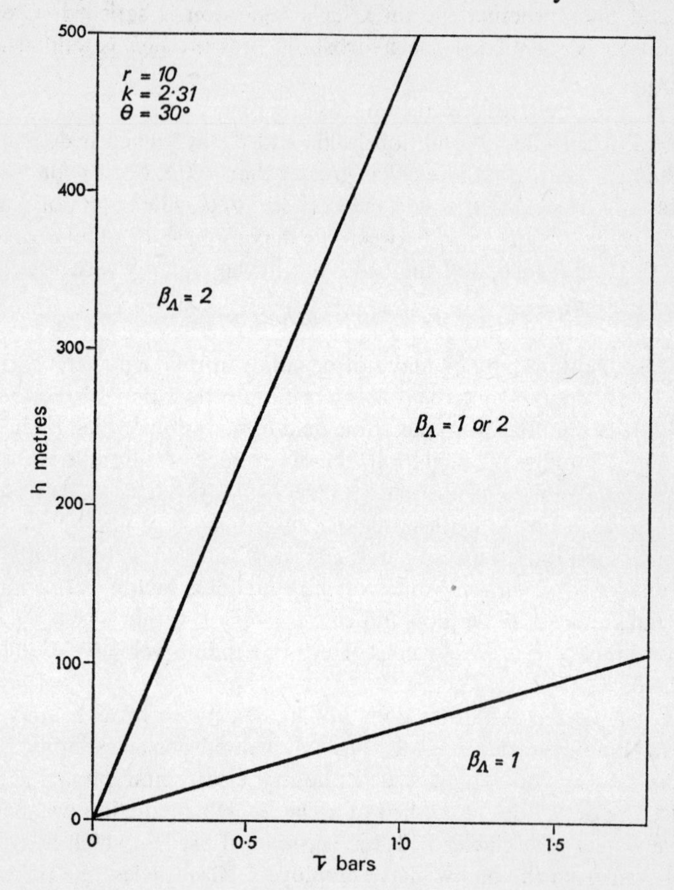

Fig. 13.4 Diagram of glacier thickness h and shear stress τ showing the regions in which $\beta_\Lambda = 1$, $\beta_\Lambda = 2$ and $\beta_\Lambda = 1$ or 2. The roughness factor $r = 10$ and $k = 2{\cdot}31$. The ratio of maximum angle θ between slope of the surface of the obstacle and the average slope of the bed is assumed to be 30°

from the obstacle increases the farther downstream one goes. A simple calculation shows that the probability is of the order of 1 at the distance Λr^2. Therefore the length of the cavity cannot exceed Λr^2.

It is evident that when the length of the cavities becomes as long as Λr^2 the bottom of the glacier touches the bed only at the tops of the obstacles. This is the condition fundamental to Lliboutry's theory of sliding (1959). I presented an argument (Weertman, 1962) that this condition can occur only for extremely rapid sliding velocities such as would exist during the avalanching of thin ice slabs. The argument

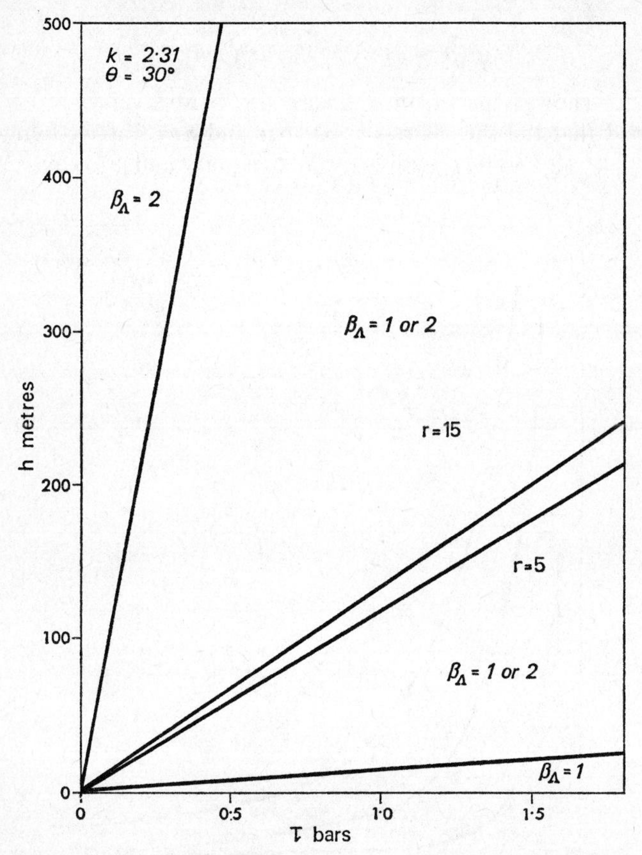

Fig. 13.5 Same diagram as Fig. 13.4 except that r = 5 and r = 15

that we presented still is valid. If an appreciable fraction of a glacier loses contact with the bed through the formation of cavities the effective hydrostatic pressure at the bottom of the glacier is increased over the value ρgh by the factor μ (the area of the bed divided by the area of ice in contact with the bed). This increase in the hydrostatic pressure increases the rate of closure of a cavity. The length of a cavity now is of the order of $\Lambda(\tau r^2/k)^n/(\mu\rho gh)^n$.

The value of μ may be estimated easily. An area $(r\Lambda)^2$ of the bed contains one obstacle of size Λ. The area of the glacier bed underneath the cavity behind this obstacle is approximately $\Lambda^2(\tau r^2/k)^n/(\mu\rho gh)^n$. From these two areas the following equation is found for the ratio μ:

$$\mu^2(\mu - 1) = (\tau r^2/k)^n/r^2(\rho gh)^n. \tag{10}$$

Fig. 13.6 shows a plot of μ against h for various values of r. It is assumed that the shear stress $\tau = 1$ bar and $n = 3$. (According to Figs. 13.4 and 13.5 it is possible for β_Λ to equal 2 and μ to equal 1 for

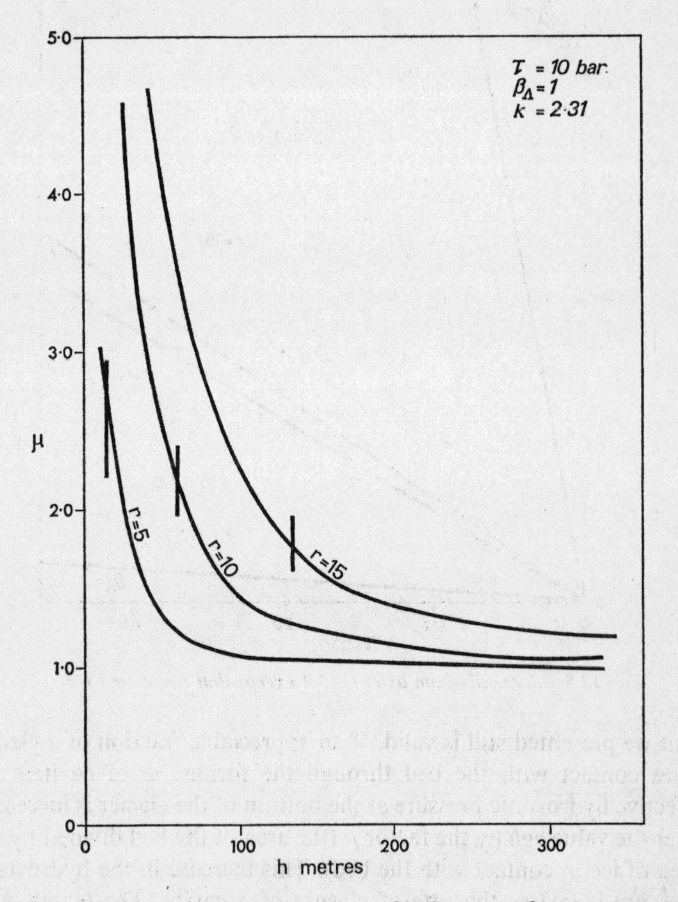

Fig. 13.6 Plot of the ratio μ (ratio of area of bed to area of ice in contact with bed) against the thickness h of a glacier when $\tau = 1$ bar, $k = 2{\cdot}31$ and $\beta_\Lambda = 1$

any ice thickness to the right of the vertical hatches in Fig. 13.6.) If the ice is riding on top of the obstacles, as it is pictured doing in Lliboutry's theory, μ is approximately equal to r^2. In order for μ to have this value under a stress of $\tau = 1$ bar and a roughness of $r = 15$, 10 or 5, the ice thickness must be less than 10 m. This conclusion is in harmony with my previous discussion (Weertman, 1962) of Lliboutry's paper. It can be seen from Fig. 13.6 that an appreciable separation of ice from rock can occur, although not to the extent envisaged by Lliboutry, for glacier thicknesses of the order of 100 m.

In Lliboutry's theory the ice separation at the bed profoundly influences the sliding velocity. It is to be emphasized strongly that the separations occurring in the present analysis do not have this strong influence on the velocity except when $\mu \approx r^2$. In fact so long as $\mu < r^2$ the sliding velocity is not influenced by separation. (This statement should be qualified to the extent that if μ is much larger than about 3 the effective resistance to sliding by obstacles smaller than Λ is greatly reduced. The sliding rate thus will be raised by an amount equal to the increase in velocity which results from the presence of a water layer smaller in thickness than the controlling obstacle size. This velocity increase is discussed in the following section. It is much smaller than that found in Lliboutry's theory.)

Our analysis of cavities was based on the assumption that the cavities are free of water. Lliboutry also considered cavities filled with water. There is no way to estimate the magnitude of the water pressure in a cavity from his analysis, yet a knowledge of this pressure is important since the sliding velocity he derived depends sensitively upon it.

I should like now to give an argument in favor of the idea of Lliboutry that cavities formed by obstacles normally are filled with water. Consider the cavity whose cross-sectional and top views are shown in Fig. 13.7. If the cavity does not contain water the pressure P^* of ice against rock at the periphery of the cavity will be smaller than the pressure P at the ice–rock interface a distance away from the cavity. This conclusion can be demonstrated quantitatively from Nye's theory (1953) of the closing of tunnels for the case when n of Glen's creep equation is equal to 3. (If n were smaller than about 2, P^* would be larger than P according to Nye's theory.) The water at the bottom of a glacier which exists in the ice–rock interface will always flow down a pressure gradient. Since the pressure gradient is towards the cavity a water-free cavity will become filled with water.

Suppose that an isolated cavity is filled with water. Assume that for a constant sliding velocity the cavity is in a steady-state condition. It is growing neither smaller nor larger. The water pressure at the bottom of the cavity must be identical with that under the ice. Otherwise water would flow into or out of the cavity. At the top of the cavity the water pressure could differ from the pressure in the ice by an amount which at most is of the order of $(\rho_w - \rho)g\Lambda$, where

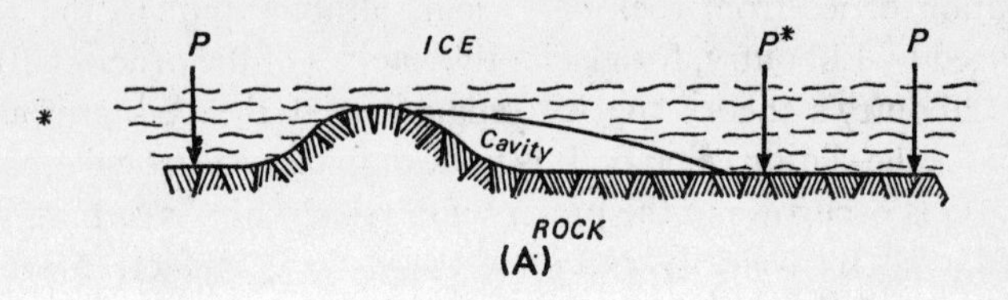

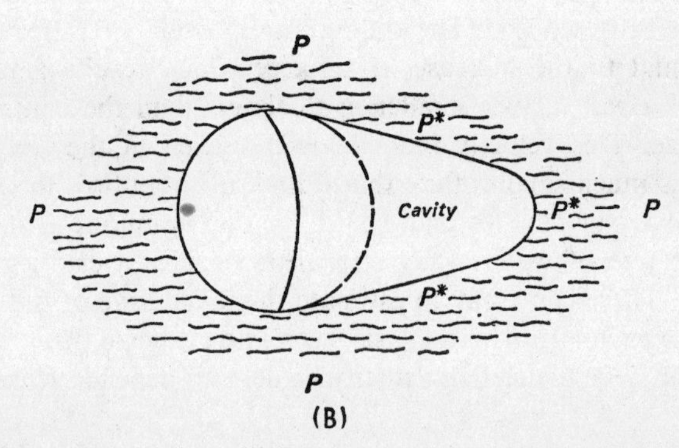

Fig. 13.7 Water-free cavity behind an obstacle: (a) cross-sectional view, (b) view from a point directly above the cavity. The pressure P is the average hydrostatic pressure at the bed and the pressure P is that exerted near the cavity (P* < P)*

ρ_w and ρ are the densities of water and ice respectively. For an obstacle 1 cm in size this pressure difference is of the order of 10^{-4} bar. Therefore only a negligibly small pressure difference is available for causing the closure of the cavity. (In a steady-state condition the closure is exactly balanced by the sliding process which opens up the cavity.) Contrary to our assumption the isolated cavity cannot be in a steady-state condition. If the cavity remains completely filled with water it must become longer and longer since there is insufficient pressure available to close it off. Eventually it will connect with other cavities. When enough cavities interconnect, the water within them can drain off. Once this happens the average diameter of a cavity will decrease through the creep flow of ice until the flow of water is so restricted that the cavity again can be completely filled with water. The cavity would have a tapered profile in the direction of ice flow similar to Fig. 13.7 except that it would not be completely closed off at its downstream end. The length of the tapered cavity would depend not only on the size of the obstacle, the roughness, and the shear stress acting across the bed, but also on the amount of water flowing at the bottom of a glacier. If the amount of water flow is large the average diameter of interconnecting cavities must be large in order to accommodate it. If it is small the average diameter will be small. In the latter situation the total area of the water-filled cavities at the glacier bed would be about the same as for the case in which the cavities were assumed to be water-free. The velocity of sliding in this situation would be the same regardless of whether the cavities were water-free or water-filled.

If the average diameter of a cavity remains large because of an abundance of meltwater supply the interconnecting water-filled cavities can be approximated in a limiting case by a sheet of water of uniform thickness at the bed of a glacier. The thickness of such a sheet of water was shown (Weertman, 1962) to be determined by the amount of water flowing through any section of the bed. We consider the effect of a water layer on the sliding velocity in the following section.

EFFECT OF THE WATER LAYER AT THE BED OF A GLACIER

The field observations of Elliston (1963) on the Gornergletscher show convincingly that water at the bottom of a glacier can change markedly the velocity of glacier movement. He finds that in the winter the glacier velocity is 20–50 per cent slower than the annual mean velocity, and the summer velocity exceeds the average by 20–80 per cent. These changes can be understood if meltwater acts as a lubricant at the bottom of a glacier. I have shown (Weertman, 1962) that if the water layer at the bottom of a glacier is thicker than the height of the controlling obstacles an increase in the sliding velocity will occur.

In our previous treatment it was found that the water layer has no effect on the sliding rate until the thickness of the water layer is as great as the height of the controlling obstacles. We should like to point out now that a water layer with a thickness an order of magnitude smaller than the controlling obstacle size can cause an appreciable increase in the sliding velocity. The reason that in the present theory a water layer of such small thickness can affect the sliding rate may be seen from equations (5) and (6). No obstacles smaller than the thickness of the water layer can cause a hindrance to the sliding motion. Thus the effective shear stress τ_i in equation (5) acting on these obstacles is zero. As a result the effective stress on the larger obstacles is raised and the sliding velocity is increased.

When the thickness d of the water layer is smaller than Λ, equation (7a) for τ_Λ becomes

$$\tau = \tau_\Lambda[1 + 2^{1/n}\{(10)^{1/n} - 1\} + \tfrac{1}{5}\{1 + \tfrac{1}{10} + (\tfrac{1}{10})^2 + \ldots (\tfrac{1}{10})^m\}]. \tag{11}$$

The number of terms in the series $1 + \frac{1}{10} + \ldots$ depends on the thickness of the water layer. If the water layer is equal to or larger than $\Lambda/10$, no term is retained in the series. If the water layer is equal to or larger than $\Lambda/100$ but smaller than $\Lambda/10$, the first term only is retained; if the layer is larger than or equal to $\Lambda/1000$ but smaller than $\Lambda/100$, the first two terms are retained; and so on. If the water layer is somewhat thicker than $\Lambda/10$ but not so thick as Λ, equation (11) reduces to

$$\tau = (2{\cdot}092)\tau_\Lambda \tag{12}$$

or $k = 2{\cdot}092$. This value of k is almost 10 per cent smaller than the previous value of 2·314. Thus the sliding velocity, which is inversely proportional to the square of k, is approximately 20 per cent larger than it would be if the water layer thickness were very much smaller. This is an appreciable increase in the sliding velocity. If the water layer thickness were of the order of $\Lambda/100$, k would be 1 per cent smaller and the sliding rate 2 per cent faster, and so on for still smaller thicknesses of the water layer.

If the water layer is thicker than the height of the controlling obstacles, equations (5) and (6) predict that the sliding velocity S^* will be

$$S^* = bB\gamma\Lambda'(\tau\gamma/\beta'k)^n r^{2n} \tag{13}$$

where β' is the value of β for the obstacles which are just bigger than the thickness of the water layer and Λ' is the size of these obstacles. The term k is equal to 1·82 if the value of β is the same for all obstacles larger than the thickness of the water layer, and $k = 2{\cdot}04$ if $\beta' = 1$ for obstacles of size Λ' and $\beta = 2$ for all larger obstacles.

Fig. 13.8 shows a plot of the ratio of the sliding velocity S^* when an appreciable water layer is present, to the ordinary sliding

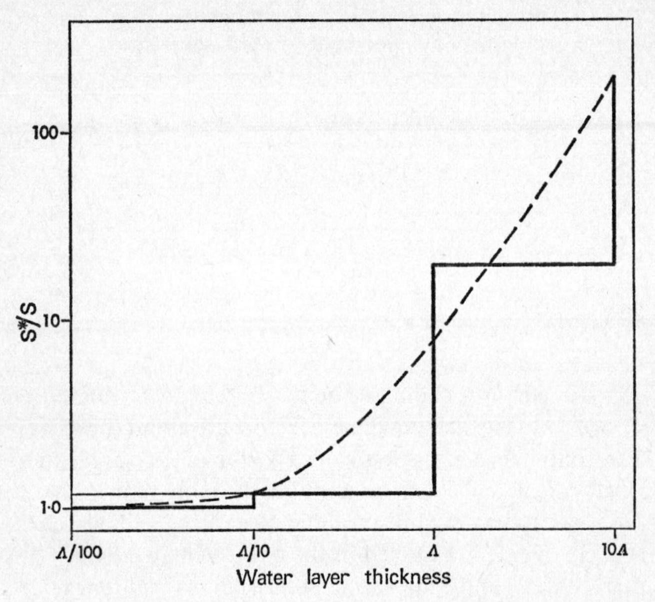

Fig. 13.8 Plot of the ratio S^*/S against thickness of the water layer at the bed. (S^* is the sliding velocity when an appreciable water layer is present and S the sliding velocity when the water layer thickness is extremely small.) The solid curve gives values of S^*/S for the bed with the discrete spectrum of obstacle sizes considered in the text. The dashed curve shows a possible variation of S^*/S with the thickness of the water layer for a bed with a continuous spectrum of obstacle sizes

velocity S versus the thickness of the water layer. (It is assumed that $\beta = 2$ for all obstacles.) The plot is a step function because we have employed a discrete rather than continuous distribution function for obstacle sizes. It is expected that the curve of S^*/S against water layer thickness for a continuous distribution function approximates to the dashed curve drawn in this figure.

DISCUSSION AND SUMMARY

In its essential features the sliding theory just presented is the same as the simpler theory developed in earlier papers (Weertman, 1957,

1958, 1962). The new values for the sliding velocity and the controlling obstacle size are approximately the same as those found previously. Table 13.1 shows the controlling obstacle sizes and roughness

Table 13.1 Controlling obstacle size and roughness factor for a sliding velocity of 80 m/yr under a 1 bar shear stress

	Older theory (Weertman, 1962)	$\beta = 1$ $k = 2{\cdot}3$	$\beta = 2$ $k = 2{\cdot}3$	$\beta_\Lambda = 1$ $k = 3{\cdot}4$
Controlling size Λ mm	1·8	3·5	6·0	3·55
Roughness factor r	16·6	14·2	18·4	17·2

factors required for a sliding velocity of 80 m a year under a shear stress τ of 1 bar. The new roughness factors are about the same as the old. The controlling obstacle size is a factor of 2–3 larger than that previously calculated. It can be seen that the more refined calculation does not lead to any significant difference in the sliding velocity or the obstacle size. It is assumed in the new calculated values that the constant factors a and b appearing in the sliding equations for S_1 and S_2 (equations (1b) and (3b)) are unity. These constant factors can be determined only by experimental tests. It seems unlikely that laboratory tests will give values for these terms which differ much from one.

Although the actual values for the sliding velocity and the controlling obstacle size have not been changed much by the new calculations, several interesting results not contained in the older theory have come to light. The most interesting of these (see dashed curve of Fig. 13.8) is the fact that a water layer smaller in thickness than the height of the controlling obstacles can change the sliding velocity by an amount of the order of 40 per cent to 100 per cent. Thus if the obstacle sizes listed in Table 13.1 are representative the changes in the flow rate observed by Elliston on the Gornergletscher could have been produced by a water layer of only 0·35–0·60 mm in thickness. This new result means that meltwater may influence the flow rate of glaciers much more than had been suspected previously.

The relaxation of the implicit assumption in the earlier version of the theory that Glen's condition always holds leads to the existence of a double-valued sliding velocity in a certain range of values of the thickness of a glacier. The two velocities in this double-valued range differ by a factor which is approximately equal to 2 or 3.

When Glen's condition does not hold, cavity formation occurs behind obstacles at the bed. Such cavities were predicted previously by Lliboutry. In Lliboutry's theory the cavity formation is so extensive that the glacier only sits on the tops of obstacles. I had previously shown (Weertman, 1962) that for a glacier of a typical thickness this condition can occur only if the glacier is sliding at velocities which are 10^7 to 10^8 larger than those actually observed. Lliboutry also considered the situation in which the glacier rested on top of obstacles as well as on top of the water trapped in the hollows of his washboard model of a glacier bed. This situation of his would correspond to the case considered in this paper in which the water layer at the bed of a glacier is sufficiently thick to have an influence on the rate of sliding. Contrary to what Lliboutry concluded we do not find that the pressure of water in the water layer has any influence on the sliding velocity.

Lliboutry's theory actually represents an extreme limiting case of the analysis of cavity formation developed in this paper. This extreme case is not likely to occur in nature unless the ice thickness is small (less than 10 m), or the ice velocity is extremely large, as would occur in avalanching ice slabs, or the shear stress at the bottom of an ice mass is considerably larger than 1 bar. Lliboutry's theory is probably valid at the snout of a glacier where the ice thickness becomes less than 10 m. Under the thinner parts of a glacier (less than 100 m in thickness) cavities should form and of course have been observed to exist.

The theory of the sliding of glaciers that has been developed in this paper is more general than that which I presented in the past. That earlier theory as well as Lliboutry's theory are contained as special cases in this more general theory.

REFERENCES

BRIGGS, L. J. (1950) 'Limiting negative pressure of water', *Journal of Applied Physics*, XXI 7, pp. 721–2.

ELLISTON, G. R. (1963) in discussion following J. WEERTMAN, 'Catastrophic glacier advances', *Bulletin of the International Association of Scientific Hydrology*, VIII 2, pp. 65–6.

Fisher, J. C. (1950) 'The limiting hydrostatic tension of water near 0°C', *Journal of Applied Physics*, xxi 10, p. 1068.

Goldthwait, R. P. (1960) 'Study of ice cliff in Nunatarssuaq, Greenland', *U.S. Snow, Ice and Permafrost Research Establishment. Technical Report* 39.

Lliboutry, L. (1959) 'Une théorie du frottement du glacier sur son lit', *Annales de Géophysique*, xv 2, pp. 250–65.

Lliboutry, L., and Brepson, R. (1963) 'Le viscosimètre à glace de Grenoble', *Union Géodésique et Géophysique Internationale. Association Internationale d'Hydrologie Scientifique. Berkeley Conference 1963. Commission des Neiges et des Glaces*, pp. 138–43.

Weertman, J. (1958) 'Transport of boulders by glaciers and ice sheets', *Bulletin de l'Association Internationale d'Hydrologie Scientifique*, x 44.

Bibliography

AGASSIZ, L. (1840) *Étude sur les glaciers* (Neuchâtel).

ALLEN, C. R., KAMB, W. B., MEIER, M. F., and SHARP, R. P. (1960) 'Structure of the Lower Blue Glacier, Washington', *J. Geol.*, LXVIII 601–25.

BATTEY, M. H. (1960) 'Geological factors in the development of Veslgjuvbotn and Vesl-Skautbotn', in W. V. Lewis (ed.), *Norwegian Cirque Glaciers, R. geogr. Soc. Res. Ser.*, IV 5–10.

BATTLE, W. R. B. (1952) 'Contributions to the glaciology of north-east Greenland, 1948–9, in Tyrolerdal and on Clavering Ø', *Meddel. Grønland*, CXXXVI.

BATTLE, W. R. B. (1960) 'Temperature observations in bergschrunds and their relationship to frost-shattering', in W. V. Lewis (ed.), *Norwegian Cirque Glaciers, R. geogr. Soc. Res. Ser.*, IV 83–95.

BATTLE, W. R. B. and LEWIS, W. V. (1951) 'Temperature observations in bergschrunds', *J. Geol.*, LIX 537–45.

BENSON, W. N. (1935) 'Notes on the geographical features of south-western New Zealand', *Geogr. J.*, LXXXVI 393–401.

BONNEY, T. G. (1876) 'Some notes on glaciers', *Geol. Mag.*, III 197–9.

BOWMAN, I. (1920) *The Andes of Southern Peru* (London).

BOYÉ, M. (1952) 'Névés et érosion glaciaires', *Rev. Géomorph. dyn.*, I 20–36.

BRIGGS, L. J. (1950) 'Limiting negative pressure of water', *J. appl. Phys.*, XXI 721–2.

CAROL, H. (1947) 'Formation of roches moutonnées', *J. Glaciol.*, I 57–9.

CHAMBERLIN, T. C. (1888) 'The rock scorings of the great ice invasions', *7th A. Rep. U.S. geol. Surv.*, 155–248.

CHAMBERLIN, T. C. and R. T. (1911) 'Certain phases of glacial erosion', *J. Geol.*, XIX 193–216.

CHARLESWORTH, J. K. (1957) *The Quaternary Era* (London), 2 vols.

CLARK, J. M., and LEWIS, W. V. (1951) 'Rotational movement in cirque and valley glaciers', *J. Geol.*, LIX 546–66.

CLAYTON, K. M. (1965) See pages 173–87 of this volume.

COLLET, L. W. (1916) 'Le charriage des alluvions dans certains cours d'eau de la Suisse', *Annls. Suisses d'Hydrographie*, IX.

COLLET, L. W. (1925) *Les Lacs* (Paris).

COTTON, C. A. (1942) *Climatic Accidents in Landscape-making* (Christchurch, N.Z.).

DAVIS, W. M. (1900) See pages 38–69 of this volume.

DAVIS, W. M. (1912) *Die erklärende Beschreibung der Landformen* (Leipzig).

DEMOREST, M. (1939) 'Glacial movement and erosion', *Am. J. Sci.*, CCXXXVII 594–605.

DURHAM, F. (1958) 'Location of the Valley Heads moraine near Tully Center determined by preglacial divide', *Bull. geol. Soc. Am.*, LXIX 1319–21.

ELLISTON, G. R. (1963) in discussion following J. WEERTMAN, 'Catastrophic glacier advances', *Bull. Ass. int. Hydrol. Sci.*, VIII 65–6.

EMBLETON, C., and KING, C. A. M. (1968) *Glacial and Periglacial Geomorphology* (London).

EVANS, J. W. (1913) 'The wearing down of rocks', *Proc. Geol. Ass. Lond.*, XXIV 241–300.

FAIRCHILD, H. L. (1907) 'Drumlins of central western New York State', *Bull. N.Y. St. Mus.*, CXI 391–443.

FAIRCHILD, H. L. (1925) 'The Susquehanna River in New York and evolution of western New York drainage', *Bull. N.Y. St. Mus.*, CCLVI 36.

FARADAY, M. (1860) 'Note on regelation', *Proc. R. Soc.*, X 440–50.

FENNEMAN, N. M. (1938) *Physiography of Eastern United States* (New York).

FISHER, J. C. (1950) 'The limiting hydrostatic tension of water near 0°C', *J. appl. Phys.*, XXI 1068.

FISHER, J. E. (1953) 'Two tunnels in cold ice at 4000 m on the Breithorn', *J. Glaciol.*, II 513–20.

FISHER, J. E. (1955) 'Internal temperatures of a cold glacier and conclusions therefrom', *J. Glaciol.*, II 583–91.

FLINT, R. F. (1957) *Glacial and Pleistocene Geology* (New York).

FORBES, J. D. (1842) 'Account of his recent observations on glaciers', *Edinb. New phil. J.*, XXXIII 338–52.

FORBES, J. D. (1843) *Travels Through the Alps of Savoy* (Edinburgh). See pages 21–37 of this volume.

GEIKIE, J. (1898) *Earth Sculpture* (London).

GERRARD, J. A. F., PERUTZ, M. F., and ROCH, A. (1952) 'Measurement of the velocity distribution along a vertical line through a glacier', *Proc. R. Soc.*, ser. A, CCXIII 546–58.

GIBSON, G. R., and DYSON, J. L. (1939) 'Grinnell Glacier, Glacier National Park, Montana', *Bull. geol. Soc. Am.*, L 681–96.

GILBERT, G. K. (1904) 'Crest lines in the High Sierras', *J. Geol.*, XII 579–88.

GILBERT, G. K. (1906) See pages 79–91 of this volume.

GILBERT, G. K. (1910) *Harriman Alaska Expedition*, vol. III: *Glaciers and Glaciation.*

GLEN, J. W. (1952) 'Experiments on the deformation of ice', *J. Glaciol.*, II 111–14.

GLEN, J. W. (1953) 'Rate of flow of polycrystalline ice', *Nature, Lond.*, CLXXII 721.

GLEN, J. W. (1955) 'The creep of polycrystalline ice', *Proc. R. Soc.*, ser. A, CCXXVIII 519–38.

GLEN, J. W. (1958) 'Measurement of the slip of a glacier past its side wall', *J. Glaciol.*, III 188–93.

GLEN, J. W. (1961) 'Measurement of the strain of a glacier snout', *Commn Neiges et Glaces, Ass. Int. Hydrol. Scient.* (Helsinki, 1960) 562–7.

GLEN, J. W., and LEWIS, W. V. (1961) See pages 188–204 of this volume.

GOLDTHWAIT, R. P. (1960) 'Study of ice cliff in Nunatarssuaq, Greenland', *Tech. Rep. 39, U.S. Snow, Ice Permafrost Res. Est.*

GOODCHILD, J. G. (1888–9) 'The history of the Eden and some rivers adjacent', *Trans. Cumberland and Westmorland Ass.*, XIV 73–90.

GROOM, G. E. (1959) 'Niche glaciers in Bünsow Land, Vestspitzbergen', *J. Glaciol.*, III 368–76.

GUTENBERG, B., BUWALDA, J. P., and SHARP, R. P. (1956) 'Seismic explorations on the floor of Yosemite valley, California', *Bull. geol. Soc. Am.*, LXVII 1051–78.

HAEFELI, R. (1940) 'Zur Mechanik aussergewöhnlicher Gletscherschwankungen', *Schweiz. Bauztg.*, CXV 179–84.

HAEFELI, R. (1951) 'Some observations on glacier flow', *J. Glaciol.*, I 496–500.

HAEFELI, R., and KASSER, P. (1951) 'Geschwindigkeitverhältnisse und Verformungen in einem Eisstollen', *Gen. Ass. int. Un. Géod., Brussels*, I 222–36.

HEIM, A. (1878) *Mechanismus der Gebirgsbildung* (Basel).

HEIM, A. (1879) 'Über die Erosion im Gebiete der Reuss', *Jb. schweizer Alpklub*, XIV 371–405.

HEIM, A. (1885) *Handbuch der Gletscherkunde* (Stuttgart).

HELLAND, A. (1882) 'Om Islands jökler og om jökelelvenes vandmaengde og slamgehalt', *Arch. Math. Nat.*, VII.

HJULSTRÖM, F. (1935) 'Studies of the morphological activity of rivers as illustrated by the River Fyris', *Bull. Geol. Inst. Uppsala*, XXV.

HOBBS, W. H. (1911) *Characteristics of Existing Glaciers* (New York).

HOLMES, C. D. (1937) 'Glacial erosion in a dissected plateau', *Am. J. Sci.*, XXXIII 217–32.

HOLMES, C. D. (1952) 'Drift dispersion in west-central New York', *Bull. geol. Soc. Am.*, LXIII 993–1010.

HOLTEDAHL, O. (1929) 'On the geology and physiography of some Antarctic and Sub-Antarctic islands', *Scientific Results, Norwegian Antartic Expedition*, III 146–68.

HOPKINS, W. (1848) 'On the elevation and denudation of the district of the Lakes of Cumberland and Westmorland', *Q. J. geol. Soc. Lond.*, IV 70.

HUGI, G. J. (1842) *Über das Wesen der Gletscher* (Stuttgart).

JACKSON, J. E., and THOMAS, E. (1960) 'Surveys and ice movements of Veslgjuvbreen', in W. V. Lewis (ed.), *Norwegian Cirque Glaciers, R. geogr. Soc. Res. Ser.*, IV 63–7.

JOHNSON, D. W. (1941) 'The function of meltwater in cirque formation', *J. Geomorph.*, IV 253–62.

JOHNSON, W. D. (1904) See pages 70–9 of this volume.

KAMB, W. B., and LACHAPELLE, E. (1964) See pages 229–43 of this volume.

KAMB, W. B., and SHREVE, R. L. (1963) 'Structure of ice at depth in a temperate glacier', *Trans. Am. geophys. Un.*, XLIV 103.

KINGERY, W. D. (1960) 'Regelation, surface diffusion and ice sintering', *J. appl. Phys.*, XXXI 833–8.

KOECHLIN, R. (1944) *Les glaciers et leur mécanisme* (Lausanne).

LACHAPELLE, E. R. (1959) 'Annual mass and energy exchange on the Blue Glacier', *J. geophys. Res.*, LXIV 443–9.

LEWIS, W. V. (1938) 'A meltwater hypothesis of cirque formation', *Geol. Mag.*, LXXV 249–65.

LEWIS, W. V. (1940) 'The function of meltwater in cirque formation', *Geogr. Rev.*, XXX 64–83.

LEWIS, W. V. (1947) 'Valley steps and glacial valley erosion', *Trans. Inst. Br. Geogr.*, XIV 19–44.

LEWIS, W. V. (1949) 'Glacial movement by rotational slipping', in *Glaciers and Climate, Geogr. Annlr., Stockh.*, XXXI 146–58.

LEWIS, W. V. (1953) 'Tunnel through a glacier', *The Times Sci. Rev.*, IX 10–13.

Lewis, W. V. (1954) 'Pressure release and glacial erosion', *J. Glaciol.*, II 417–22.

Linton, D. L. (1940) 'Some aspects of the evolution of the rivers Earn and Tay', *Scott. geogr. Mag.*, LVI 1–11, 69–79.

Linton, D. L. (1949*a*) 'Unglaciated areas in Scandinavia and Great Britain', *Irish Geogr.*, II 25–33, 77–9.

Linton, D. L. (1949*b*) 'Watershed breaching by ice in Scotland', *Trans. Inst. Br. Geogr.*, XVII 1–16.

Linton, D. L. (1955) 'The problem of tors', *Geogr. J.*, CXXI 470–87.

Linton, D. L. (1957) 'Morphological contrasts of eastern and western Scotland', in R. Miller and J. W. Watson (eds.), *Geographical Essays in Memory of Alan G. Ogilvie* (Edinburgh).

Linton, D. L. (1957) See pages 130–48 of this volume.

Linton, D. L. (1962) 'Glacial erosion on soft-rock outcrops in central Scotland', *Biul. Peryglac.*, XI 247–57.

Linton, D. L. (1963) See pages 149–72 of this volume.

Ljunger, E. (1930) 'Spaltentektonik und Morphologie der Schwedischen Skaggerrack-Küste', *Bull. geol. Instn. Univ. Uppsala*, XXI 1–478.

Lliboutry, L. (1959) 'Une théorie du frottement du glacier sur son lit', *Ann. Géophys.*, XV 250–65.

Lliboutry, L. (1965) *Traité de Glaciologie* (Paris).

Lliboutry, L. (1968) 'General theory of subglacial cavitation and sliding of temperate glaciers', *J. Glaciol.*, VII 21–58.

Lliboutry, L., and Brepson, R. (1963) 'Le viscosimètre á glace de Grenoble', *Commn Neiges et Glaces, Ass. Int. Hydrol. Scient.* (Berkeley, 1963) 138–43.

McCabe, L. H. (1939) 'Nivation and corrie erosion in W. Spitsbergen', *Geogr. J.*, XCIV 447–65.

McCall, J. G. (1952) 'The internal structure of a cirque glacier', *J. Glaciol.*, II 122–30.

McCall, J. G. (1954) 'Glacial tunnelling and related observations', *Polar Rec.*, VII 120–36.

McCall, J. G. (1960) See pages 205–28 of this volume.

MacClintock, P., and Apfel, E. T. (1944) 'Correlation of drifts of the Salamanca re-entrant, New York', *Bull. geol. Soc. Am.*, LV 114–16.

Marr, J. E. (1906) 'The influence of the geological structure of the English Lakeland upon its present features', *Q. J. geol. Soc. Lond.*, LXII lxvu–cxviii.

MARTONNE, E. DE (1911) 'L'érosion glaciaire et la formation des vallées alpines', *Ann. Géogr.*, XIX 289–317; XX 1–29.

MATHEWS, W. H. (1959) 'Vertical distribution of velocity in Salmon glacier, British Columbia', *J. Glaciol.*, III 448–54.

MATTHES, F. E. (1900) 'Glacial sculpture of the Bighorn Mts., Wyoming', *21st A. Rep. U.S. geol. Surv.*, 167–90.

MATTHES, F. E. (1930) See pages 92–118 of this volume.

MATTHES, F. E. (1949) 'Glaciers', in O. E. Meinzer (ed.), *Physics of the Earth*, vol. IX: *Hydrology*, 149–220 (New York).

MEIER, M. F. (1960) 'Mode of flow of Saskatchewan Glacier, Alberta, Canada', *U.S. geol. Surv. Prof. Pap.* 351.

MILL, H. R. (1895) 'Bathymetrical survey of the English Lakes', *Geogr. J.*, VI 46–73, 135–66.

MONNETT, V. E. (1924) 'The Finger Lakes of Central New York', *Am. J. Sci.*, VIII 33–5.

MOSS, J. H., and RITTER, D. F. (1962) 'New evidence regarding the Binghamton sub-stage in the region between the Finger Lakes and the Catskills', *Am. J. Sci.*, CCLX (2) 81–106.

MUIR, J. (1916) 'Studies in the Sierra: mountain sculpture; origin of Yosemite valleys; glacial denudation', *Bull. Sierra Club*, X 68 ff.

MULLER, E. H. (1963) 'Reduction of preglacial erosion surfaces as a measure of the effectiveness of glacial erosion', *Rep. 6th Int. Congr. Quaternary Ass.* (Warsaw) III 233–43.

NYE, J. F. (1952) 'A comparison of the theoretical and measured long profile of the Unteraar Glacier', *J. Glaciol.*, II 103–7.

NYE, J. F. (1952) 'The mechanics of glacier flow', *J. Glaciol.*, II 82–93.

NYE, J. F. (1953) 'The flow law of ice', *Proc. R. Soc.*, ser A, CCXIX 477–89.

NYE, J. F. (1960) 'The response of glaciers and ice-sheets to seasonal and climatic changes', *Proc. R. Soc.*, ser A, CCLVI 559–84.

NYE, J. F. (1963) 'Theory of glacier variations', in W. D. Kingery (ed.), *Ice and Snow; Properties, Processes and Applications* (Proceedings of a conference held at M.I.T., Cambridge, Mass., 1962) 151–61.

PARTHASARATHY, A., and BLYTH, F. G. H. (1959) 'The superficial deposits of the buried valley of the river Devon near Alva, Clackmannan, Scotland', *Proc. Geol. Ass.*, LXX 33–50.

PERUTZ, M. F. (1950) 'Direct measurement of the velocity distribution in a vertical profile through a glacier', *J. Glaciol.*, I 382–3.

PILLEWIZER, W. (1950) 'Bewegungsstudien an Gletschern der Jostedalsbre in Südnorwegen', *Erdkunde*, IV 201–6.

PLAYFAIR, J. (1802) *Illustrations of the Huttonian Theory of the Earth* (Edinburgh).

REID, H. F. (1896) 'The mechanics of glaciers', *J. Geol.*, IV 912–28.

RICHTER, E. (1896) 'Geomorphologische Beobachtungen aus Norwegen', *Sitz. Wiener Akad. Math.-Naturw.*, CV 147–89.

RUSSELL, I. C. (1895) 'Influence of debris on the flow of glaciers', *J. Geol.*, III 823–32.

RÜTIMEYER, L. (1869) *Über Thal- und See-Bildung* (Basel).

SAUSSURE, H. B. DE (1779–96) *Voyages dans les Alpes* (Paris), 4 vols.

SAVAGE, J. C., and PATERSON, W. S. B. (1963) 'Borehole measurements in the Athabasca glacier', *J. geophys. Res.*, LXVIII 4521–36.

SCHROEDER, H. (1928) *Erläuterungen zur geologischen Karte von Preussen, Blatt Halberstadt.*

SHARP, R. P. (1954) 'Glacier flow: a review', *Bull. geol. Soc. Am.*, LXV 821–38.

SHREVE, R. G. (1961) 'The borehole experiment on Blue Glacier, Washington', *Commn Neiges et Glaces, Ass. int. Hydrol. Scient.* (Helsinki 1960) 530–1.

SOONS, J. M. (1959) 'The sub-drift surface of the lower Devon valley', *Trans. geol. Soc. Glasg.*, XXIV 1–7.

STREIFF-BECKER, R. (1938–9) 'Zur Dynamik des Firneises', *Z. Gletscherk., Berl.*, XXVI 1–21.

STRØM, K. M. (1949) 'The geomorphology of Norway', *Geogr. J.*, CXII 19–27.

TARR, R. S. (1905) 'Drainage features of Central New York', *Bull. geol. Soc. Am.*, XVI 229–42.

TAYLOR, G. (1914) 'Physiography and glacial geology of east Antarctica', *Geogr. J.*, XLIV 365–82, 452–67, 553–71.

TAYLOR, G. (1922) 'The physiography of the McMurdo Sound and Granite Harbour region', *British Antarctic Expedition 1910–13* (London).

THOMPSON, J. (1860) 'On recent theories and experiments regarding ice at or near its melting point', *Proc. R. Soc.*, X 152–60.

THORARINSSON, S. (1939) See pages 119–29 of this volume.

TYNDALL, J. (1858) 'On some physical properties of ice', *Phil. Trans. R. Soc.*, CXLVIII 211–29.

VON ENGELN, O. D. (1937) 'Rock sculpture by glaciers', *Geogr. Rev.*, XXVII 478–82.

von Engeln, O. D. (1961) *The Finger Lakes Region* (Ithaca, N.Y.).

Waeber, M. (1942–3) 'Observations faites au glacier de Tré-la-tête', *Rev. Géogr. alp.*, xxx/xxxi 319–43.

Washburn, H. B., and Goldthwait, R. P. (1937) 'Movement of South Crillon Glacier, Crillon Lake, Alaska', *Bull. geol. Soc. Am.*, xlviii 1653–63.

Weertman, J. (1957) 'On the sliding of glaciers', *J. Glaciol.*, iii 33–8.

Weertman, J. (1958) 'Transport of boulders by glaciers and ice sheets', *Bull. Ass. int. Hydrol. sci.*, x 44.

Weertman, J. (1962) 'Catastrophic glacier advances', *Commn Neiges et Glaces, Ass. int. Hydrol. Scient.* (Obergurgl, 1962) 31–9.

Weertman, J. (1964) See pages 244–68 of this volume.

Weertman, J. (1966) 'Effect of a basal water layer on the dimensions of ice-sheets', *J. Glaciol.*, vi 191–207.

Weertman, J. (1967) 'An examination of the Lliboutry theory of glacier sliding', *J. Glaciol.*, vi 489–94.

Williams, H. S., *et al.* (1909) 'Description of the Watkins Glen–Catatonk district, New York', *U.S. Geologic Atlas*, folio 169.

Wooldridge, S. W. D., and Linton, D. L. (1955) *Structure, Surface and Drainage in South-east England* (London).

Wright, C. S., and Priestley, R. (1922) 'Glaciology', in *Results of British (Terra Nova) Antarctic Expedition of 1910–13* (London).

Index

REV. DR PAUL D. VROLIJK, Ph.D (2008) in Philosophy, University of Bristol (through Trinity College Bristol) is an ordained minister in the Church of England. Currently he is the Chaplain of Aquitaine in the Anglican Diocese in Europe.